FIREFLY

SECOND EDITION

PRACTICAL ASTRONOMY

STORM DUNLOP

FIREFLY BOOKS

A FIREFLY BOOK

Published by Firefly Books Ltd. 2012

First printing

Publisher Cataloguing-in-Publication Data (U.S.)

Dunlop, Storm.
Practical astronomy / Storm Dunlop.
2nd ed.
[208] p. : col. ill., col. maps ; cm.
Includes bibliographical references and index.
Summary: The basics for amateur astronomers, including how to observe, whether with the naked eye, binoculars or a small telescope, and where and when to look; start by finding constellations and visible planets, and locating more challenging phenomena.
ISBN-13: 978–1–77085–143–6 (pbk.)
1. Astronomy — Observers' manuals. 2. Astronomy — Popular works.
I. Title.
523 dc23   QB64.D84   2012

Library and Archives Canada Cataloguing in Publication

A CIP record for this title is available from Library and Archives Canada

Published in the United States by
Firefly Books (U.S.) Inc.
P.O. Box 1338, Ellicott Station
Buffalo, New York 14205

Published in Canada by
Firefly Books Ltd.
66 Leek Crescent
Richmond Hill, Ontario L4B 1H1

This book was developed by Philip's,
a division of Octopus Publishing Group Ltd,
Endeavour House, 189 Shaftesbury Avenue,
London WC2H 8JY

Front Cover: *(clockwise from top left)* eclipsed Moon *(H. J. P. Arnold)*, star map *(Wil Tirion)*, star trails *(Chris Cook)*, Rho Ophiuchi Dark Cloud *(Stephen Pitt)*. In the foreground is a 100 mm refractor telescope.
Back Cover: *(above)* star map *(Wil Tirion)*, *(below)* cross-section through prismatic binoculars.
Title page: *The Lagoon Nebula (M8) in the constellation of Sagittarius.*

Printed in China

CONTENTS

INTRODUCING ASTRONOMY

EXPLORING THE SKY

BEGINNING ASTRONOMY

Astronomy is a fascinating hobby that anyone can follow. You do not need to be – as some people seem to imagine – "mathematically minded" to start, or even to become a very experienced observer. Yet astronomy is one of the few hobbies where not only can you gain great enjoyment, but if you want to, you can very easily make observations of great scientific value.

What is perhaps even more surprising is that you do not need complicated equipment – or indeed any equipment at all. So if you are a beginner, do not feel that you must rush out and buy the most expensive telescope that you can afford. This could be a big mistake, as it might prove to be completely unsuitable for the objects which you later find are most interesting. If you must buy anything, a pair of binoculars is certainly far more useful at first, but even these are not essential, and some of the things that may be observed with the naked eye are described on page 20. Similar lists of objects and observations that you can make with binoculars and telescopes are given on pages 54 and 55, while details of how to choose (and test) binoculars and telescopes are also given later.

Because this book is intended for amateurs who are just starting to make practical observations, it does not attempt to discuss the equipment or methods used in some of the more specialized fields of study, including some where highly experienced amateurs are making major contributions to astronomical science. Some of these areas are mentioned, and some examples of results are included in the images that are shown, but for detailed discussion the reader is referred to the various sources (such as books, Internet sites, and societies) listed at the end of this work.

How to use this book

There is such a range of objects in the sky that there is always something to see, from meteors and aurorae in the Earth's atmosphere, to more distant planets and stars, and extending out to remote galaxies far off in space. There is a lot of pleasure to be gained from "rambling" around the sky, looking at whatever objects happen to be available at the time, or which take your fancy. Everyone has to start by learning to find their way around, and by recognizing the different constellations. This is the way that this book begins, as well as by giving general information on how to set about observing.

After a while, most astronomers find that they become particularly interested in a few classes of objects, on which they tend to concentrate their attention. Because these may require different types of equipment, or different methods of observation, they are individually described in the various chapters later in this book.

Beginners often find it a bit confusing because there are so many different objects, each of which is best observed in a particular way. Similarly, when first starting it is not always easy to know what may be seen or studied with particular equipment. A number of flow charts and tables have therefore been given, which should help you to find the relevant sections where the different subjects are discussed, and to move on to the next stage in discovering the fascination of astronomy.

Beginners

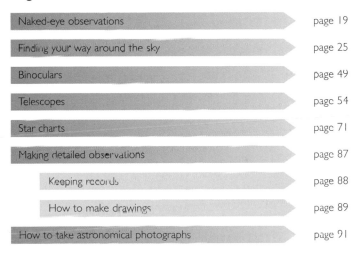

Starting to observe

There are a few things to remember when you start to observe, but first make sure that you are warm and dry – no one can observe properly if they are uncomfortable. Even in summer it may get quite cold at night, so wear plenty of clothing. A quarter of all body heat is lost through the head, so a hat is often essential. Dampness (especially underfoot) makes the problem of keeping warm much worse, so a dry site is better than standing on wet grass. Stone and concrete may become very cold, and hard to the feet during a long observing session, so wooden duckboards which provide some insulation are ideal.

Try to pick a spot that also offers some protection from the wind, not only because it will be warmer, but also because the wind can shake binoculars or a telescope, making viewing more difficult. Even a simple windbreak can help a lot. Dampness as it affects equipment is discussed later in this chapter. Observers in warmer locations have other problems and may find that mosquito repellant is an essential part of their equipment.

The eyepieces of many telescopes can assume awkward positions and heights at times, so you may need some form of steps. These must be sturdy and stable, but reasonably easy to move. A stout wooden box may be a satisfactory alternative. Diagonals can help to make the eyepiece more accessible, but they introduce a mirror-image reversal, which means that the view may be difficult to compare with charts or photographs. Looking high overhead is easier if you use a reclining, garden chair (preferably one with arms) rather than craning your neck – and it is also far more comfortable. Some observers like to lie flat on the ground on a sheet of thick foam or an airbed.

It also helps to have everything to hand. Some telescope tripods incorporate space for small items such as eyepieces, but a garden table is more suitable for all the bits and pieces that you may want.

Eyesight

The pupil of the eye responds almost instantaneously to major changes in light by expanding or contracting, but true dark adaptation takes place when a pigment (known as "visual purple") builds up inside the retina. This takes about 30 minutes or more, during which time the eyes slowly become more sensitive. It helps if the eyes are protected from bright lights before you go out to observe – some observers even wear sunglasses before starting an observing session. There are strong indications that dark adaptation becomes better and faster (within limits) the more frequently it is used, and this is another argument for trying to observe as often as possible.

Bright light quickly destroys dark adaptation at any time – even viewing the Moon through a telescope will do this – but a very dim red light has least effect, so make sure that you have one for examining charts and writing notes. Cover a suitable lamp or pen-light with red paper or plastic, and either change the bulb to a dimmer one, or make sure that the covering lets through only a weak light. When fully dark adapted, and under a clear sky, there is a surprising amount of light from the stars alone, often more than enough to enable you to move around without too much difficulty.

The advantage of binocular observing is that you use both eyes at once, in the normal relaxed manner. With a telescope, try to conquer the natural tendency to close the "unwanted" eye, because this only leads to strain on both. With practice one eye can be "ignored," but if this proves too difficult, or if there is a lot of stray light causing interference, wear an eyepatch that allows you to keep both eyes open.

The most troublesome eyesight defect is astigmatism, which can cause stellar images to appear elongated or misshapen. Long- or short-sight does not pose many problems, because most binoculars and telescopes have sufficient range of focusing adjustment for this to be accommodated. If spectacles have to be worn all the time, take particular care in selecting equipment. Some suggestions about this are given later.

At first most beginners wonder if there is something wrong with their eyesight when they cannot see faint planetary detail, or pick out the dimmer stars. But it is surprising how quickly one's perception improves with practice, so the more frequently you can observe the better. Experienced observers frequently use averted vision – looking slightly to one side of the faint object they want to see, so that the image falls on a more sensitive part of the retina. This does work, although exact positions may become a little more difficult to judge. While telescopes and binoculars should be kept as rigid as possible, tapping the eyepiece very lightly may sometimes bring faint stars into view, because the eye is extremely sensitive to movement.

Where to observe

A dark observing site is most important. Interference from lights prevents the proper dark adaptation that is so important for seeing faint objects. However, naked-eye and binocular observers have an advantage in that they can move around more freely than anyone with a telescope. Observing from within the shadow of a wall or building can make a great deal of difference. In towns and cities, light pollution is very bad and often only part of the sky can be seen, restricting the objects that are visible (Fig. 1.1). Taking a portable telescope out into the country may be one answer, but with perseverance, much observing can be carried out even under poor conditions. It may be an advantage when learning the constellations if only the brighter stars can be seen through city lights. A considerable number of deep-sky objects (which are

► *Fig 1.1 The constellation of Perseus under light-polluted and clear skies. Poor conditions may still be used to find one's way around the sky.*

generally the most severely affected by light pollution) can still be seen with a moderately sized telescope, even under poor conditions.

Light pollution has, unfortunately, become extremely bad in some countries and regions. Luckily, both national and local authorities are becoming aware of the problem and – partly thanks to lobbying by the International Dark Sky Association and similar local bodies – are starting to take action. It is to be hoped that many more countries will follow the splendid example set by the Czech Republic and pass national legislation to counter light pollution. Lighting manufacturers, too, are now offering lights with excellent horizontal cut-off, so that the light is sent downward (where it is required) and not up into the sky.

It is possible to combat light pollution to a certain extent by the use of filters that block specific wavelengths of light, such as the lines produced by certain sodium lighting. But most sky glow over towns and cities consists of such a mixture of light that this is only partially successful. More troublesome for many astronomers are the local security lights installed by neighbors. These send a wash of blinding light – which has been called "light trespass" – across neighboring homes and gardens when a cat walks within range of their infrared sensors. There is little that can be done about these lights, except trying to persuade the owners to turn them downward so that the light is cast on to the ground and is not emitted horizontally. It is also sometimes possible to find a corner of your garden that is in shadow.

When to observe

Not all astronomy is done at night. Apart from the study of the Sun, which needs special techniques for the sake of safety, it is sometimes of advantage to study Venus in the daylight, when the contrast between its brilliance and the sky is reduced, and faint details are easier to see. In some observational fields it is important to try to make observations as soon as possible after the Sun has set, or immediately before it rises – searching for comets and novae are just two examples.

Calculating the time of sunrise or sunset, depending as it does upon one's position on Earth, is too complex to describe here. Because of the effects of refraction (Fig. 1.2), when the Sun appears to be on the

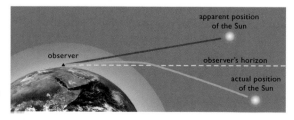

apparent position
of the Sun

observer

observer's horizon

actual position
of the Sun

◀ *Fig. 1.2 Curved light-paths produced by refraction in the atmosphere cause all astronomical objects to appear higher in the sky.*

▶ Fig 1.3 At low latitudes (top: south of the equator) the Sun sets at a steep angle and astronomical twilight occurs every night, unlike the conditions in summer farther toward the poles (bottom: northern hemisphere).

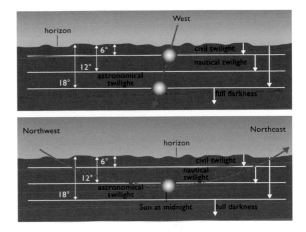

horizon it is actually below it by about 35 minutes of arc (approximately the same as its diameter). Newspapers and diaries often give the time of sunset and "lighting-up" time, at the end of civil twilight, when the Sun is 6° below the horizon, as well as the corresponding times in the morning. Although a useful guide, these are local standard or summer times and are correct for only a few observers. More important to astronomers is the length of astronomical twilight, which occurs while the Sun is between 12–18° below the horizon. It lasts at least 70 minutes after sunset, beginning the same time before sunrise (Fig. 1.3). At moderately high latitudes – in fact beyond 48.5° N and S – astronomical twilight persists all night during some part of the summer months.

Figure 1.5 shows how sunrise and sunset times, and the duration of twilight and full night vary throughout the year for four different latitudes. An astronomical yearbook will give precise details of how long twilight lasts at your latitude on any particular date.

Observers at high latitudes, however, have some compensations for the long twilight, because it is in summer that noctilucent clouds are likely to be seen. They are also most favored with aurorae. In addition, it is sometimes more difficult to see objects near the Sun if observers are situated close to the equator, because the objects set more rapidly (Fig. 1.6).

Moonlight also causes considerable interference with many types of observation, mainly because the scattered light increases

▲ Fig. 1.4 Observing the Moon and Venus in daylight reduces extreme contrasts.

the brightness of the background sky, reducing the contrast between it and the faint light of galaxies and similar objects. It may mean, for example, that meteor showers may be well-nigh unobservable in some years if they occur at the time of Full Moon. For those not interested in the Moon, a total lunar eclipse can offer the chance of snatching a few valuable observations that would otherwise be unobtainable.

Observations of superior planets – those outside the orbit of the Earth – and of the minor planets (generally called the asteroids) are usually best undertaken when they are close to opposition (page 46), crossing the meridian at around midnight. Opposition is also the time when they are closest to the Earth, and offer the largest disk sizes. Mercury and Venus, the inferior planets, are best placed at elongation (page 46), when they are half-illuminated. Naturally, observations are also undertaken at other times, and with some objects, such as comets, there can be no choice of best observing period. The most favorable conditions for stars, clusters, nebulae, and galaxies are when they cross the meridian at midnight and are highest in the sky.

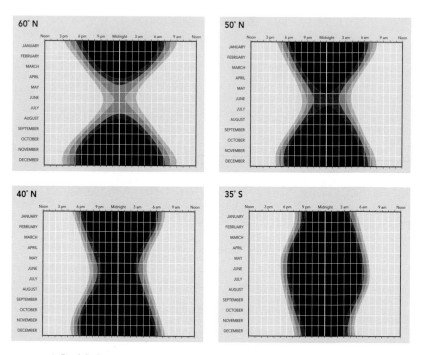

▲ Fig. 1.5 Sunrise and sunset times, and the duration of twilight and full night, vary throughout the year and according to the observer's latitude. Shown here are the times for 60°N, 50°N, 40°N, and 35°S.

▶ Fig. 1.6 At the equator (top) an object A, close to the Sun, is well below the horizon by the time astronomical twilight begins. At high northern latitudes (bottom) the object is easily observable the same length of time after sunset.

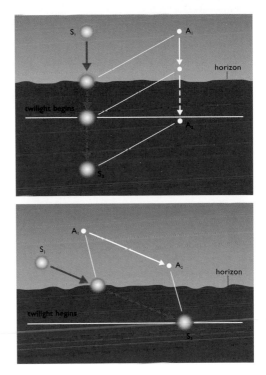

Atmosphere and seeing

Not all nights are equally suitable for observing, and the overall conditions (or seeing) are rated on a scale devised by Antoniadi, a famous planetary observer (Table 1.1). Much depends upon meteorological conditions, including those high overhead, but very local effects caused by the observatory, telescope, and observer are also involved. When there is strong turbulence, variations in the density of the air layers refract light and produce scintillation (Fig. 1.7). This gives rise to random movements of the images and changes in their brightness. Planetary disks appear blurred, and if stars are close to the horizon, where refraction disperses the light, they may show marked changes in color – one of the causes of many so-called "UFO" reports. In a telescope, the effects are more pronounced and images wander around, and go in and out of focus. Under such conditions it may be necessary to persevere, awaiting the moments when the seeing steadies, although this may not happen very often during the course of a night. Naturally, photography and serious observing may well be impossible under extreme conditions, which frequently occur on cold nights, even when the air near the ground appears to be calm.

TABLE 1.1: ANTONIADI SCALE OF SEEING	
I	Perfect seeing, without a quiver
II	Slight undulations, with periods of calm lasting several seconds
III	Moderate seeing, with larger air tremors
IV	Poor seeing, with constant troublesome undulations
V	Very bad seeing, scarcely allowing a rough sketch to be made

The tube currents found in some reflectors, and air turbulence within an observatory have the same general effect. If equipment (particularly telescopes) is not kept in an unheated observatory or store it should be allowed to reach the same temperature as the outside air before observing begins, to help to prevent these problems.

There is always absorption in the atmosphere – it is sometimes called atmospheric extinction – decreasing the brightness of astronomical objects. This absorption is at a maximum close to the horizon, and decreases toward the zenith. General haziness from dust or pollution (especially downwind of a large, or industrial, city) degrades the seeing even more. Absorption may often be a problem with naked-eye observations, particularly estimating the magnitudes of variable stars and meteors at low altitudes, so take extra care under such conditions. In general, observations of even the brighter stars and planets are difficult or impossible within 10° of the horizon.

Large bodies of water do have a stabilizing effect upon local temperatures and conditions, and can noticeably improve seeing conditions. A slightly damp and even slightly hazy atmosphere can give rise to superb, steady conditions, and these also occur after a band of rain has passed by, bringing clearer air behind it. Although a damp haze may sometimes appear unfavorable, users of binoculars and telescopes often find that they can "see through it" and experience good viewing conditions. Although patchy cloud-cover may be infuriating, it can frequently bring good conditions between the clouds. In general,

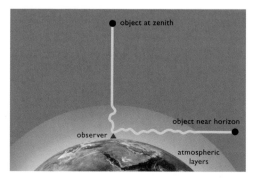

◄ Fig 1.7 Objects close to the horizon show greater scintillation than those near the zenith as the light has a longer path through the dense, fluctuating, lower layers of the atmosphere.

cumulus clouds which build up during the day have a tendency to die away and disperse after dark, but layer clouds, such as those associated with depressions, usually persist into the night. Some of the best seeing conditions come with the passage of cold fronts, even though the air behind may still contain a fair amount of cloud.

Dampness can also be a problem when it produces condensation, or dewing, on telescopes, binoculars and other items, if they become colder than the surrounding air, or are taken into a warm atmosphere. Dewcaps should always be used, and objectives and mirrors must be covered before being taken indoors. If a glass surface does become dewed, it should not be wiped, with the risk of damage to the optical coating, but the dampness may be dispersed by fanning with a piece of paper.

Essential equipment

The only items that are essential are a red light, and a notebook and pen or pencil. (A pencil is always worth having anyway, as some ball-point pens refuse to write if it is cold.) Try to keep a note of what you observe – even if it is only to record that some object was, or was not, visible. It is a good habit, too, to enter the date and time of every observation, so a watch or clock is also needed, preferably one set to Universal Time (see page 86) to prevent confusion. Try to make little sketches of planets, lunar features, galaxies, or anything else that takes your fancy. They do not have to be great works of art, but just good enough to give an impression of what you can see. Gradually, as you build up your knowledge of the sky, your notes and drawings will become better, more comprehensive, and probably more specialized. It may seem like a chore, but actually they will quickly add to your enjoyment of the hobby. (Keeping records is more fully described on page 88.)

What equipment do you possess?

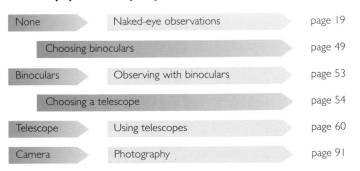

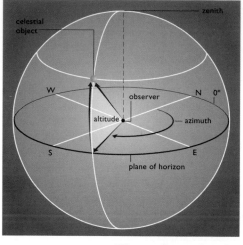

◀ Fig 1.8 *Altitude is always measured upward from the plane of the horizon, and azimuth around from the north point toward the east.*

The celestial sphere

The stars and all other celestial bodies, such as the Sun and Moon, appear to be located on the inside of a vast sphere, centered on the observer, and rotating toward the west. Although we know that this view – which of course was held by the ancients – is not true, it is still a useful way of thinking about the sky. Just as latitude and longitude are used to locate positions on the surface of the Earth, a system of celestial coordinates is used on the celestial sphere, and is described later (page 72). The north and south celestial poles are extensions of the Earth's rotational axis, and the celestial equator is in line with the Earth's equator.

Of more immediate use is the horizontal coordinate system, which is centered on the observer's own position (Fig. 1.8). Positions are specified in terms of altitude and azimuth, from which comes the alternative name of the altazimuth system. Altitude is the elevation in degrees above (+) or, occasionally, below (−) the horizon. Azimuth is the angle in degrees measured from the north point (0°), through east (90°), south (180°), and west (270°) back to north (360°/0°). The point directly overhead (at +90°) is known as the zenith, and that directly beneath the observer's feet (at −90°) is the nadir. When stars are invisible, as at twilight or during full daylight, altitude and azimuth may be the only means of specifying the position of certain objects, such as a bright fireball.

The observer's position on Earth

Exactly which part of the celestial sphere is ever visible depends on the observer's position on the Earth. At the North Pole only the northern portion of the sky – and thus half of the stars – may be seen, and of course a similar situation applies at the South Pole. In both cases every star in that half of the sky is visible (at least theoretically) whenever the Sun is below the horizon, and the heavens rotate about the corresponding celestial pole, which is found at the zenith. The bright star Polaris (α Ursae Minoris) is very close to the true position of the northern celestial pole, but unfortunately in the south the pole is not marked by any conspicuous star.

At other latitudes, stars from both northern and southern celestial hemispheres may be seen. At 45° north, for example, Polaris appears halfway down toward the northern horizon, and many of the southern stars are visible. Now, however, only the stars within 45° of the celestial pole are circumpolar, remaining above the horizon all the time, and are visible on any clear night (Figs. 1.9 and 1.10). The remaining stars rise and set, and those that are visible throughout the course of a night change slowly with the seasons. In theory, anyone at the equator could see both poles and all the stars in the sky (although only half at once), but in practice the effects of refraction and absorption complicate the issue.

For any observer the most important imaginary line in the sky is the meridian, which is a great circle running right round the sky through the north and south points of the horizon, the north and south celestial poles, and the observer's zenith and nadir (Fig. 1.8). It may be regarded as the celestial equivalent of the observer's meridian of

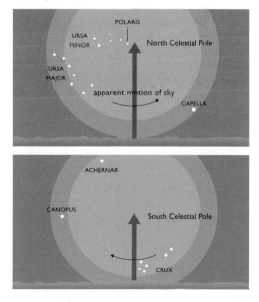

▶ Figs 1.9 and 1.10
The altitude of the poles is always equal to the observer's latitude. At 45°N (above), Capella is just circumpolar, but at 35°S (below), Canopus sets for part of the night and Crux brushes the horizon.

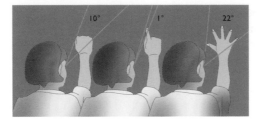

◀ Fig 1.11 Angles may be estimated approximately by using a hand at arm's length, the results being correct for nearly everyone.

longitude on the surface of the Earth. When objects cross this north–south line they are said to transit the meridian. (Transit telescopes, fixed to observe just this line in the sky, were once important equipment at every observatory.) An object culminates, reaching its highest altitude in the sky, as it transits the meridian. Circumpolar stars, of course, cross the meridian both above and below the pole, and these events are known as upper and lower culmination, respectively.

Because the sky appears as a sphere, centered on the observer, all distances between objects may be expressed as angles, 360° forming a complete circle. It is frequently useful to be able to make these measurements, even if only approximately. Figure 1.11 shows how various angles may be estimated by using the hand held out at arm's length. Another method is to use a rule, graduated in centimeters. When held at arm's length, one centimeter is approximately equal to 1 degree. If the

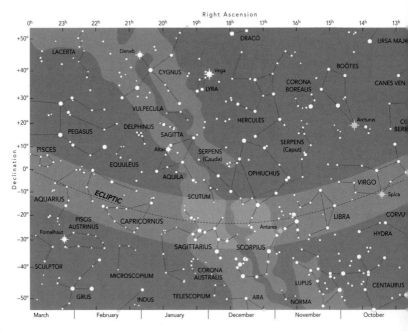

location of any object needs to be specified exactly with respect to the observer's horizon, altitude and azimuth are employed. The proper instrument for measuring both of these coordinates at the same time is an astrocompass, but various types of orienteering compass will allow you to measure elevation (that is, altitude) as well as providing azimuth. If your compass does not allow you to measure vertical angles, a simple device may be made from a straight edge, a protractor, and a weighted line fixed to the center of the protractor. This will measure angles to a sufficient degree of accuracy for most purposes.

Changes throughout the year

The Earth's true rotation period, measured with respect to the stars (the sidereal day), is almost exactly $23^h 56^m 04^s$; this is about four minutes shorter than the Sun's average, apparent rotation period (the mean solar day), because of the Earth's motion around the Sun. As a result, when measured by ordinary civil time, individual stars rise (and set) about four minutes earlier each day, slowly shifting westward across the night sky. At times they may come too close to the Sun to be visible, thus giving rise to unavoidable seasonal gaps in observation.

▼ Fig. 1.12 The band of the zodiacal constellations in which the Moon and the planets may be found is centered on the ecliptic, the Sun's apparent path.

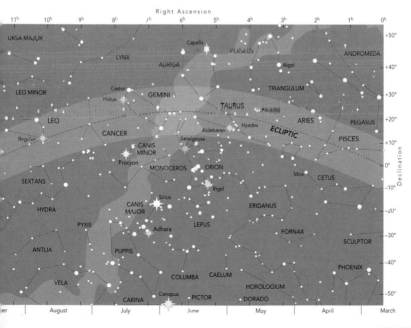

We know that stars actually move in space, and change their positions relative to one another. However, their distances from us are so great that any changes caused by this proper motion take many centuries to become apparent to the naked eye. For most purposes we can imagine that the Sun, Moon, planets, and other bodies move against this "fixed" background. The Sun appears to trace out a path known as the ecliptic (Fig. 1.13), completing the circuit of 360° in one year. Because the Earth's axis is tilted by just under 23.5°, the ecliptic makes the same angle with the equators of the Earth and of the celestial sphere. During the course of the year this gives rise to the changing elevation of the Sun, and to the seasons.

The Moon and major planets follow paths which usually lie within about 8° of the ecliptic. In ancient times there were twelve constellations in this band (about 16° wide), and it was known as the zodiac (Fig. 1.12). Astrologers divided the zodiac into twelve equal portions, each 30° long, even though the constellations were of different sizes, and regarded their "star signs" as having special significance. With the passing of the centuries, and the effects of precession, the position of the ecliptic has altered with respect to the background stars. The Sun and planets now pass through several constellations that were not included in the original zodiac, such as Ophiuchus and Orion, both of which are ancient constellations. It is small wonder then that astronomers regard astrology and "star signs" as sheer superstition. Only one of the symbols for the zodiacal constellations is in common use by astronomers, that for Aries (♈), which is used to indicate one of the two important points on the celestial sphere where the ecliptic crosses the celestial equator.

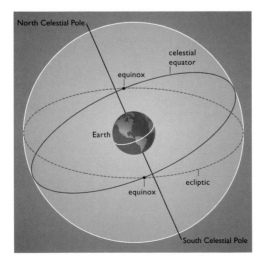

◀ Fig. 1.13 On the celestial sphere the poles and the equator are directly related to those of the Earth. As the Earth rotates the whole sky passes across any observer's meridian during the course of a day.

LEARNING YOUR WAY

Naked-eye observing

There are many things that you can do with just the naked eye. Most important is learning the constellations and how to find your way around the sky, as described shortly. Despite the introduction of "GOTO" telescopes, which will automatically point to any object held in their internal database, or to any point specified by celestial coordinates, this skill remains essential for any serious astronomer, and provides excellent practice for later binocular and telescopic observing. The same applies to many individual objects. The Moon, for example, shows about the same amount of detail to the naked eye as most of the planets do through small telescopes, so making drawings of the lunar surface as it appears with no optical aid can be a very useful exercise.

Many stars are distinctly colored and several of the open and globular clusters are visible, as are a few of the external galaxies. The star clouds of our own Galaxy appear as the Milky Way, which runs right around the celestial sphere. Its beauty can only be truly appreciated with the naked eye, observing in dark, unpolluted skies. You can trace its dark clouds without too much difficulty, particularly the Great Rift which runs down from Cygnus toward the galactic center. Any

▲ *Fig. 2.1 Trying to spot the thin crescent Moon is a challenge to the naked-eye observer. Here the Moon is just over one day old. Venus is on the left.*

TABLE 2.1: NAKED-EYE OBJECTS		
Const.	Desig.	Name and remarks
And	M31	Great Andromeda Galaxy
Cnc	M44	Praesepe – open cluster
CVn	M3	Globular cluster
Cen	ω	Fine globular cluster
Cru	–	Coalsack – dark nebula
Dor/Men	LMC	Large Magellanic Cloud – nearest galaxy
Her	M13	Globular cluster
Lyr	ε	Double to very good eyesight
Ori	M42	Orion Nebula
Per	h & χ	The Double Cluster – twin open clusters
Per	M34	Open cluster
Tau	θ	Easy double star
Tau	M45	Pleiades – finest open cluster
Tri	M33	Galaxy, visible only under optimum conditions
Tuc	SMC	Small Magellanic Cloud
Tuc	47	Globular cluster – NGC 104
UMa	ζ	Mizar – wide double with Alcor

instruments give too high a magnification for the Milky Way's full extent to be seen, so that only wide-field photographs do it any justice. Frequently, too, the tails of comets can only be distinguished without any optical aid at all (or else only with very specialized equipment), because they are so faint and of such low contrast.

It is a test for the eyesight to see if some stars appear double, and there are some variable stars – apart from occasional novae – which either rise above the naked-eye limit, or can be seen all the time. Spotting the thinnest crescent of the New Moon is also a challenge (Fig. 2.1), and it is supposed to be possible to see four satellites of Jupiter with good eyesight, under very favorable conditions.

The major planets, Venus, Mars, Jupiter, and Saturn, may usually be readily recognized. Mercury, being always close to the Sun, is often more difficult to identify, and both the large but distant planet Uranus and the minor planet Vesta are near the limit of naked-eye visibility, even when closest and under good seeing conditions. They may be located with the use of suitable charts. Using just the naked eye, the movements of planets and comets can be followed over a period of time and plotted against the stars.

Perhaps the most important naked-eye observations that can be made, however, are those of meteors and aurorae. Efficient observation of the former, in particular, requires a good knowledge of the constellations.

Learning the constellations

If you are interested in astronomy it is essential to be able to find your way around the sky, and it is best to begin by learning some of the most important groups of stars, or constellations. There are 88 of these and their names form a rather odd mixture because some date back to antiquity and commemorate mythological beings and creatures, and others are of far more recent origin, frequently describing scientific instruments. Many different names have been suggested over the years, but most have been discarded. The patterns of stars rarely bear even the slightest resemblance to the objects after which they have been named, and in general the individual stars are at greatly different distances and are quite unrelated to one another, so that they merely appear to be close together.

Both the constellation boundaries and the official names were only finally settled by international agreement in 1930, and some of the older names are still occasionally encountered. The Quadrantid meteor stream, for example, is named after the former constellation of Quadrans Muralis (the Mural Quadrant, an early astronomical instrument). In addition to its proper, Latin name (often derived from earlier Greek words), nearly every constellation has a common name, frequently just a translation from the Latin. Most astronomers use the Latin names, and these are given in Table 2.2, together with some of the more common, colloquial names that are occasionally found, especially in older books. If you are learning the constellations for the first time, try to use the Latin names, even though they may seem a little more difficult to remember or to pronounce, because they are internationally known and are found in all the most useful charts and catalogs. Do not be put off by the thought that 88 different constellations and names may be too many to learn, because as we have seen earlier, the part of the sky that is visible varies with the observer's position on Earth, and also depends upon the season and the time of night. Many constellations may therefore be permanently, seasonally or temporarily invisible. In any case it would be quite easy to learn one new constellation every night.

Once the major constellations are known and recognized, you will find that the fainter ones are soon distinguished. Occasionally one or more planets may appear in a constellation, making the star patterns difficult to recognize – at least at first glance. This can only happen to the constellations along the ecliptic, and the planets are usually easy to recognize from their appearance, and because they show less tendency to scintillation than neighboring stars. Their motions over a period of time also serve to identify them, and these are discussed later in this chapter.

TABLE 2.2: CONSTELLATIONS			
Name	Genitive	Abbreviation	Common name
Andromeda	Andromedae	And	Andromeda
Antlia	Antliae	Ant	Air Pump
Apus	Apodis	Aps	Bird of Paradise
Aquarius	Aquarii	Aqr	Water Bearer
Aquila	Aquilae	Aql	Eagle
Ara	Arae	Ara	Altar
Aries	Arietis	Ari	Ram
Auriga	Aurigae	Aur	Charioteer
Boötes	Boötis	Boo	Herdsman
Caelum	Caeli	Cae	Chisel
Camelopardalis	Camelopardalis	Cam	Giraffe
Cancer	Cancri	Cnc	Crab
Canes Venatici	Canum Venaticorum	CVn	Hunting Dogs
Canis Major	Canis Majoris	CMa	Great Dog
Canis Minor	Canis Minoris	CMi	Little Dog
Capricornus	Capricorni	Cap	Sea Goat
Carina	Carinae	Car	Keel (of a ship)
Cassiopeia	Cassiopeiae	Cas	Cassiopeia
Centaurus	Centauri	Cen	Centaur
Cepheus	Cephei	Cep	Cepheus
Cetus	Ceti	Cet	Whale
Chamaeleon	Chamaeleontis	Cha	Chameleon
Circinus	Circini	Cir	Compass
Columba	Columbae	Col	Dove
Coma Berenices	Comae Berenices	Com	Berenice's Hair
Corona Australis	Coronae Australis	CrA	Southern Crown
Corona Borealis	Coronae Borealis	CrB	Northern Crown
Corvus	Corvi	Crv	Crow
Crater	Crateris	Crt	Cup
Crux	Crucis	Cru	Southern Cross
Cygnus	Cygni	Cyg	Swan
Delphinus	Delphini	Del	Dolphin
Dorado	Doradus	Dor	Dorado
Draco	Draconis	Dra	Dragon
Equuleus	Equulei	Equ	Foal
Eridanus	Eridani	Eri	River Eridanus
Fornax	Fornacis	For	Furnace
Gemini	Geminorum	Gem	Twins
Grus	Gruis	Gru	Crane
Hercules	Herculis	Her	Hercules
Horologium	Horologii	Hor	Pendulum Clock
Hydra	Hydrae	Hya	Water Snake
Hydrus	Hydri	Hyi	Lesser Water Snake
Indus	Indi	Ind	Indian
Lacerta	Lacertae	Lac	Lizard

TABLE 2.2: CONSTELLATIONS (cont.)			
Name	Genitive	Abbreviation	Common name
Leo	Leonis	Leo	Lion
Leo Minor	Leonis Minoris	LMi	Little Lion
Lepus	Leporis	Lep	Hare
Libra	Librae	Lib	Scales
Lupus	Lupi	Lup	Wolf
Lynx	Lyncis	Lyn	Lynx
Lyra	Lyrae	Lyr	Lyre
Mensa	Mensae	Men	Table (Mountain)
Microscopium	Microscopii	Mic	Microscope
Monoceros	Monocerotis	Mon	Unicorn
Musca	Muscae	Mus	Fly
Norma	Normae	Nor	Level (square)
Octans	Octantis	Oct	Octant
Ophiuchus	Ophiuchi	Oph	Serpent Bearer
Orion	Orionis	Ori	Orion
Pavo	Pavonis	Pav	Peacock
Pegasus	Pegasi	Peg	Pegasus (winged horse)
Perseus	Persei	Per	Perseus
Phoenix	Phoenicis	Phe	Phoenix
Pictor	Pictoris	Pic	Easel
Pisces	Piscium	Psc	Fishes
Piscis Austrinus	Piscis Austrini	PsA	Southern Fish
Puppis	Puppis	Pup	Stern (of a ship)
Pyxis	Pyxidis	Pyx	Compass
Reticulum	Reticuli	Ret	Net
Sagitta	Sagittae	Sge	Arrow
Sagittarius	Sagittarii	Sgr	Archer
Scorpius	Scorpii	Sco	Scorpion
Sculptor	Sculptoris	Scl	Sculptor
Scutum	Scuti	Sct	Shield
Serpens	Serpentis	Ser	Serpent
Serpens caput			Serpent's head
Serpens cauda			Serpent's tail
Sextans	Sextantis	Sex	Sextant
Taurus	Tauri	Tau	Bull
Telescopium	Telescopii	Tel	Telescope
Triangulum	Trianguli	Tri	Triangle
Triangulum Australe	Trianguli Australis	TrA	Southern Triangle
Tucana	Tucanae	Tuc	Toucan
Ursa Major	Ursae Majoris	UMa	Great Bear
Ursa Minor	Ursae Minoris	UMi	Little Bear
Vela	Velorum	Vel	Sails (of a ship)
Virgo	Virginis	Vir	Virgin
Volans	Volantis	Vol	Flying Fish
Vulpecula	Vulpeculae	Vul	Little Fox

Stars

Most of the brightest stars have individual names, many of which were given by Arab astronomers in the Middle Ages. These tend to be a bit confusing (as similar names apply to different stars), as well as being awkward to remember, or seemingly well-nigh impossible to pronounce. However, astronomers very rarely use these old names nowadays, except in a very few, particularly important cases, preferring to use the Greek-letter designations given by the German astronomer Bayer at the beginning of the 17th century. Bayer took each constellation in turn, generally calling the brightest star Alpha (α), the next brightest Beta (β), the third Gamma (γ), and so on down toward fainter stars, and through the alphabet. This system has been retained because it is so convenient, even though it only applies to the brightest stars, and despite the fact that in many cases we now know that the stars should have been arranged in a slightly different order of brightness. Various other methods of identifying the fainter stars have been used, and some of these schemes are discussed later. Greek letters are listed in Table 2.3.

The Bayer letter for an individual star is always followed by the Latin name of the constellation concerned, written in the genitive. These genitives are also given in the table, as are the standard three-letter abbreviations. The latter are nearly always used in lists of objects, and

◀ Fig. 2.2 The constellation of Orion, photographed by Akira Fujii. It is well known to observers all over the world and can be used as a pointer to other constellations. Mintaka (δ Orionis), the right-hand star in Orion's "belt," lies almost exactly on the celestial equator.

TABLE 2.3: GREEK ALPHABET					
α	alpha	β	beta	γ	gamma
δ	delta	ε	epsilon	ζ	zeta
η	eta	θ	theta	ι	iota
κ	kappa	λ	lambda	μ	mu
ν	nu	ξ	xi	ο	omicron
π	pi	ρ	rho	σ	sigma
τ	tau	υ	upsilon	φ	phi
χ	chi	ψ	psi	ω	omega

are probably easier to remember than the genitives. You will find that these, and the various other names soon become very familiar. As an example of how the system works we may take Mintaka, a star nearly on the celestial equator, in the constellation of Orion (Fig. 2.2). The name Mintaka is derived from the Arabic *Al Mintaka,* "the Belt," it being the northernmost of the three forming Orion's "belt." Bayer decided that it was fourth in importance in the constellation so called it "δ (Delta) Orionis," usually written by astronomers as "δ Ori."

The brightness of stars (or of any astronomical objects, such as planets) is measured in magnitudes. For the moment it is sufficient to note that the scale works "backward," so that the brightest stars have the smallest magnitudes. Once again this was initiated by the ancient astronomers, who regarded the brightest stars as being the most important, and therefore of the "first magnitude," the next brightest of the "second magnitude," and so on. Under good conditions the faintest stars that may be seen with the naked eye are about magnitude 6 to 6.5. The magnitude scale now has a sound scientific basis, with a first-magnitude star defined as being exactly 100 times as bright as a sixth-magnitude star. (This accounts for the fact that a difference of one magnitude equals a difference in brightness by a factor of 2.512.) The scale also has a precisely defined zero point, but a few very bright objects had to be given negative values, such as −1.4 for Sirius, the brightest star. Venus, the brightest planet, may reach magnitude −4, while the brightness of the Full Moon is about magnitude −13.

Finding your way around the sky

The star charts in this book are given in two forms. The first set, following shortly, are designed to help you to find your way around the sky and to recognize the major constellations. They emphasize the patterns formed by the brighter stars, not just those of the conventional constellations. Two charts cover the northern and southern circumpolar regions and six the equatorial band. For ease of initial identification the order of the equatorial charts is slightly different from that usually given. A later section explains the system of celestial coordinates that enables the position of any object to be stated precisely, and the second set of charts (pages 75–82) carries these coordinates. Both sets show all the constellations and stars to magnitude 5. Larger-scale charts would be needed to show, in a satisfactory manner, all the stars down to about magnitude 6, which is generally accepted as the naked-eye limit in the absence of light pollution. The individual charts in the two sets cover the same regions of the sky, so they may be directly compared with one another.

Most observation is carried out in the evening, so the charts indicate when the particular regions are on the meridian at 22:00 hours (10 p.m.) local standard time. For every 2 hours earlier (or later) that you observe, a date one month earlier (or later) will be approximately correct. A device known as a planisphere (Fig. 2.3), showing a flat projection of the sky, and with a rotating mask that can be set to any date or time, is extremely useful for showing which constellations are visible at any instant. You should obtain one that is correct for your latitude. There are also numerous computer programs that will provide an accurate view of the sky visible at any time or place.

A planisphere cannot accurately represent the motions of the planets, and their positions must be obtained in some other manner. In this book, some details are given in the tables for each planet. It used to be the case that full information had to be taken from a handbook or almanac for the year concerned, but many astronomers now use a

◄ Fig. 2.3 A planisphere shows which part of the sky is above the horizon at any date and time, so it is easy to determine which objects will be visible during the night.

computer program instead. Some programs can be linked directly to a telescope, controlling its movement, so that objects such as planets can be located and tracked automatically. Despite this automation, it is still an advantage to be able to identify the different constellations and find one's way around the sky using no more than the naked eye.

The constellation with which observers begin depends mainly on where they live on Earth. If you live in the northern hemisphere, Ursa Major (the Great Bear) is undoubtedly best, while in the southern hemisphere, Crux (the Southern Cross) is very distinctive. Orion and most other equatorial constellations are known to observers all over the world. Several prominent groups of stars (or asterisms), which are not true constellations in themselves, are also very useful for guidance in some regions of the sky.

The descriptions that follow use the Latin names of the constellations, the standard, three-letter abbreviations and occasionally the genitives. These are all given in Table 2.2 (pages 22–23). Similarly, the actual Greek-letter names of stars are shown on the charts and given in the text. Occasional mention is made of the angular distance between particular stars. Methods by which such angles can be either estimated, or else measured with reasonable accuracy using simple devices, were described earlier. It is important to remember that in these descriptions – and in dealing with any star chart – compass directions do not refer to points around the observer's normal horizon but to the celestial sphere. "North" is always toward the North Celestial Pole, and "south" away from it, even with circumpolar constellations that may appear "upside down." Similarly, looking south along the meridian between the two celestial poles, "west" is always to the right.

Finding your way around the sky

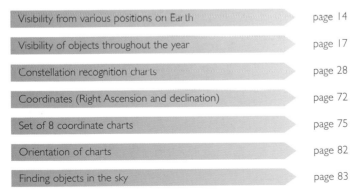

The northern polar constellations

The seven main stars of Ursa Major (UMa) form the group (or asterism) known as the "Big Dipper" in North America, and the "Plough" elsewhere, as well as by many other names. It is familiar to people who know nothing else about astronomy. The distinctive shape is easy to recognize, and it is usually visible at some time during the night, although on winter evenings it may be low on the northern horizon – and thus obscured – depending upon the latitude of where you live.

A line through the two "Pointers," β UMa (Merak) and α UMa (Dubhe), indicates the position of Polaris, the Pole Star, in Ursa Minor (UMi). The distance between the two Pointers is about 5°, and that from α UMa to Polaris is roughly 28°. Polaris (α UMi) is the one star that appears motionless in the sky. All the other stars seem to circle counterclockwise around it very slowly throughout the night. In fact it lies about 1° from the North Celestial Pole, and does trace out its own small circle, which becomes visible in long-exposure photographs. Polaris is very slightly variable, but the changes are so small that it is not suitable for amateur observation. Its visual brightness is about magnitude 2, similar to the other bright star in Ursa Minor, β UMi (Kochab).

If you imagine a line from ζ UMa (Mizar: a naked-eye, binocular and telescopic binary) through Polaris, this passes very close to the true celestial pole. When extended by about the same distance, this line points to δ Cas, in the "W"-shaped constellation of Cassiopeia (Cas), lying in the Milky Way. Because this constellation lies opposite Ursa Major, one or other constellation is usually readily visible to help with orientation on the sky. Starting at Polaris and moving counterclockwise, mentally draw a line at right-angles to the last one. This passes between the bright stars Deneb (α Cygni) and Vega (α Lyrae), slightly farther from the pole. Returning to the Big Dipper or Plough, the top of the "bowl" (δ and α UMa) points unmistakably toward the constellation of Auriga (Aur), slightly south of its brightest star, Capella (α Aur).

Key to constellation maps

Magnitudes							Ecliptic
	-1	0	1	2	3	4 5	Constellation figures
Double stars		Diffuse nebulae					Prominent groups
Variable stars		Planetary nebula					Large asterisms
Open star cluster		Galaxies					Directions
Globular star cluster		Milky Way					

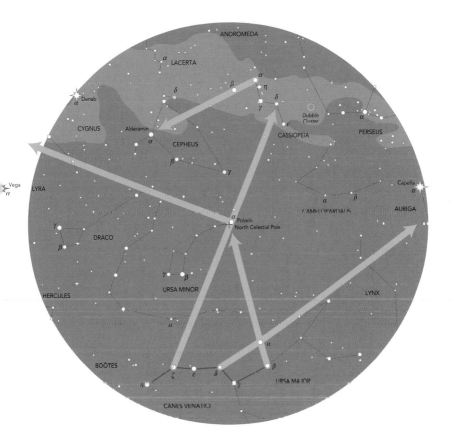

The faint constellation of Draco (Dra) straggles around the pole and Ursa Minor, its quadrilateral "head" lying northeast of Vega, and its "tail" between Polaris and Ursa Major. The unremarkable, roughly pentagonal, constellation of Cepheus (Cep) lies partly in the Milky Way between Cygnus and Cassiopeia, extending up toward the pole. The brightest star, α Cep (Alderamin), is indicated by the line from α and β Cas.

On the other side of Cassiopeia, between it and Auriga and still in the Milky Way, lies the rather more obvious constellation of Perseus, described later. The insignificant groupings of Camelopardalis (Cam) and Lynx (Lyn), and some of the fainter stars of Ursa Major lie in the large area between the seven main stars of that constellation and Auriga.

Equatorial constellations
January, February, March

Orion (Ori) is the key constellation in this region. With its distinctive shape straddling the celestial equator, it is a guide for northern and southern observers alike. The red supergiant star Betelgeuse (α Ori) and the brighter, brilliant white Rigel (β Ori) are distinctive, as is the line of three second-magnitude stars forming the "belt" of this mythical hunter. The northernmost star of the belt, Mintaka (δ Ori), lies nearly on the celestial equator. South of the belt are the stars of the "sword," with the famous Orion Nebula (M42) in the center. The nebula is faintly visible as a hazy patch even to the naked eye. Follow the line of the belt stars southward and they point approximately toward Sirius (α Canis Majoris), the brightest star in the sky, with a magnitude of −1.4. To the northwest the line of the belt passes just south of orange Aldebaran (α Tauri), with the nearby "V" of the Hyades cluster. Still farther to the west, the same line leads in the general direction of the other very distinctive cluster in Taurus, the Pleiades (M45).

The constellation of Taurus (Tau), although supposed to represent a bull, largely consists of a "head" (Aldebaran and the Hyades) and "horns," the tips of which are marked by single moderately bright stars, north of Orion, between that constellation and Auriga. Apart from these, there are a few fourth-magnitude stars to the south and west of Aldebaran.

Auriga (Aur), with bright Capella (α Aur) and the distinctive triangle of the "Kids" to the west, appears to form an irregular pentagon, although the southernmost star is actually β Tau. The Milky Way runs through the constellation, but is less distinct here than in Cygnus, or in the dense star clouds of the southern hemisphere.

Northeast of Orion lies the zodiacal constellation of Gemini (Gem) with the distinct bright pair of stars Castor (α Gem) and Pollux (β Gem). Pollux, the southernmost, is the brighter of the pair. Lines of stars running back toward Orion form the "bodies" of the "twins." A line from δ UMa through β UMa also indicates Castor and Pollux.

South of Castor and Pollux, and forming more or less an equilateral triangle with Betelgeuse and Sirius, is the isolated bright star Procyon (α CMi) in the small constellation of Canis Minor (CMi), which contains only one other bright star (β CMi), to the northwest. Canis Major (CMa), on the other hand, has several bright stars apart from Sirius, including one (ε CMa) of first magnitude.

The "spine" of Canis Major points along the Milky Way, and a right-angle turn at ζ Puppis leads (past the "two triangles" of Puppis) on to brilliant Canopus (α Carinae), the second-brightest star in the sky, magnitude −0.7. Between Canopus and Orion lie the constellations of Columba and Lepus.

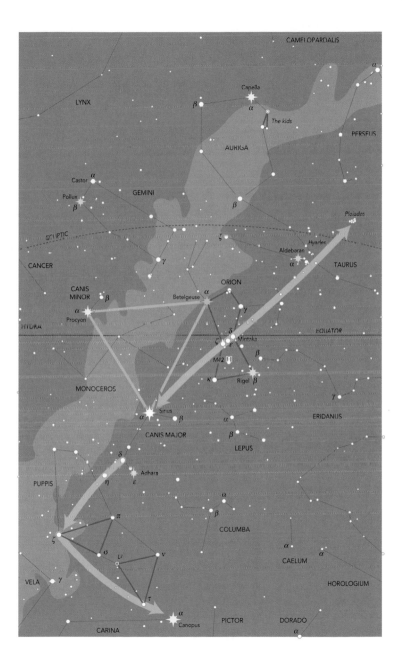

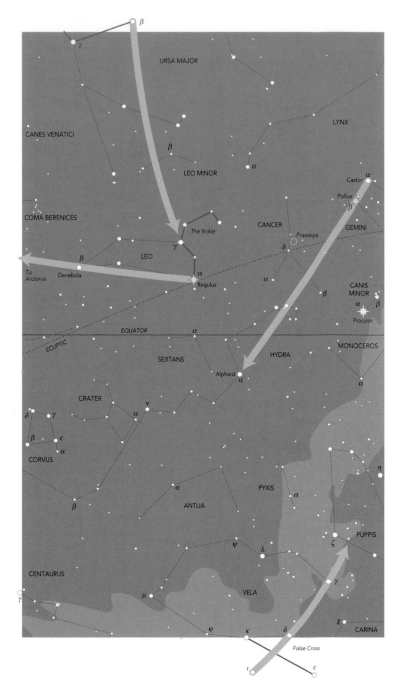

Equatorial constellations
March, April, May

South of Ursa Major (follow the line of the Pointers away from Polaris), and well to the east of the constellations of Gemini and Canis Minor, is the distinct constellation of Leo. Regulus (α Leo), the brightest star, is of first magnitude. It lies almost on the ecliptic and may sometimes be occulted by the Moon. It forms part of the well-known asterism of the "Sickle," a reversed "question mark" of stars, curving north and to the west. The main body of the constellation extends to the east toward second-magnitude Denebola (β Leo).

A line running through Regulus and Denebola is one way of finding the brightest star in the northern hemisphere, orange Arcturus (α Boötis) far off in the east. To the north of Leo, between it and Ursa Major, lie Lynx (Lyn), with one third-magnitude star, and the very faint, uninteresting constellation of Leo Minor (LMi). Another small, faint constellation, Cancer (Cnc), lies between Gemini and Leo, and is only really notable for the Praesepe open cluster (M44).

South of Cancer and east of Procyon (α CMi) is the small asterism forming the "head" of Hydra. Although actually the largest constellation in terms of area, Hydra (Hya) has only one bright star, Alphard (α Hya), southeast of Regulus. (A line through Castor and Pollux points to both the "head" and Alphard.) From there the constellation trails first south, then eastward, roughly parallel to the equator, in a string of faint stars which ends even farther east than Arcturus.

Three faint constellations lie north of Hydra. The faintest of these is Sextans – its brightest star is magnitude 4.5 – lying south of Regulus and northeast of Alphard. The next, Crater, lies to the southeast, west of the rather more conspicuous quadrilateral formed by the stars of Corvus.

There are brighter stars in the south. There the "False Cross" contains stars from Vela (Vel) and Carina (Car). The arm formed by ι Car and δ Vel points north to γ Vel, a striking double, and onward to the approximate region of ζ Puppis. Other first-magnitude stars in Vela are κ, also part of the "False Cross," and orange λ Vel to the north.

Equatorial constellations
May, June, July

If you continue the curve of the "tail" of Ursa Major round and down toward the equator for approximately 30°, you arrive at Arcturus, α Boötis (magnitude 0.0), the fourth brightest star in the sky after Sirius, Canopus, and α Centauri, all of which are in the southern hemisphere. Boötes (Boo) is fairly distinct as a "P"-shaped group of fairly bright stars, north and east of Arcturus. The brightest of these, Izar (ε Boo), is a well-known double star.

The same arc in the sky leads on from Arcturus in the general direction of Spica (α Virginis), just south of the ecliptic, and beyond that to the four main stars of Corvus. Virgo (Vir) is a large constellation on both sides of the equator, but with no stars other than Spica brighter than magnitude 3. Almost due south of Denebola in Leo lies β Vir, and the constellation can be traced in rough quadrilaterals of third- and fourth-magnitude stars eastward below Arcturus.

The constellation of Libra (Lib), once the "claws" of Scorpius to the east, is about as far south as Spica. The brightest star (β Lib) forms a triangle with Spica and Arcturus to the west. The fainter, third-magnitude α Lib is almost exactly on the ecliptic slightly farther to the southwest.

In the north, the single bright star (α) of the small constellation of Canes Venatici (CVn) is isolated in the center of the arc formed by the "tail" of Ursa Major. To the south, northeast of Denebola, lies the faint constellation of Coma Berenices (Com), which like Virgo contains many distant galaxies.

To the east of Boötes is the almost perfect circlet of stars forming Corona Borealis (CrB). This attractive constellation contains second-magnitude Gemma (α CrB) – also known as Alphekka – and the important variable R Coronae Borealis. The straggling line of stars forming Serpens Caput, half of the constellation of Serpens, leads down toward the south.

North of Crux and the two bright stars α and β Centauri are the remaining bright stars of Centaurus, forming a rough pentagon south of the "tail" of Hydra. The irregular but very approximately rectangular constellation of Lupus runs roughly southwest to northeast toward red Antares (α Scorpii).

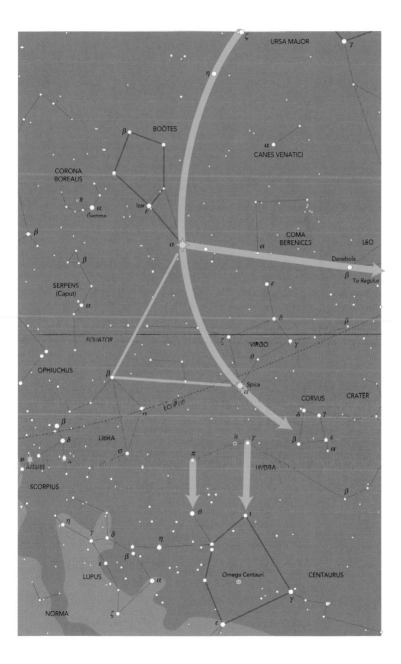

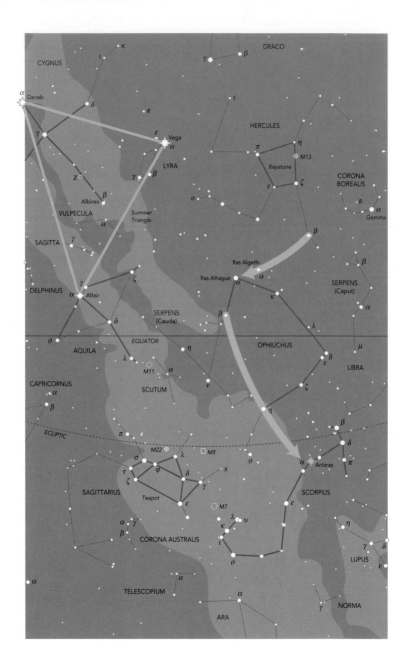

Equatorial constellations
July, August, September

This is the region dominated by the (northern) "Summer Triangle" formed by Vega (α Lyrae), Deneb (α Cygni), and Altair (α Aquilae) in the north and the stars of Scorpius (Sco) and Sagittarius (Sgr) in the south. It also contains some of the most striking parts of the Milky Way, especially in the south. The large cross formed by the stars of Cygnus (Cyg) is very distinct, its main arm from Deneb to Albireo (β Cyg, a fine telescopic double) pointing south along the "Great Rift" in the Milky Way. The roughly "T"-shaped constellation of Aquila is to the south of the Rift, and lies across the equator.

The constellation of Lyra (Lyr) mainly consists of Vega and a small parallelogram of stars to the southeast. Slightly north and east of Vega is ε Lyr, the famous "Double Double," a wide pair of stars distinguishable to good eyesight, each of which proves to be double in a telescope.

Between Lyra and Corona Borealis to the west are the four stars of the "Keystone," part of Hercules (Her), with "arms" and "legs" stretching out from each corner. Between the two western stars of the "Keystone" lies the globular cluster M13, which is visible as a hazy spot to the naked eye.

None of the stars in Hercules is very bright, only α Her (Ras Algethi) marking the "head" in the south is about third magnitude. It is a double, one star of which is variable, and lies close to the brighter star α Ophiuchi (Ras Alhague).

The constellation of Ophiuchus (Oph) – the ancient "Serpent Bearer" – sprawls across the equator and the ecliptic in a rough pentagon. It divides the two chains of stars that form the halves of Serpens: the "head" (Serpens Caput) in the west, and the "tail" (Serpens Cauda) in the east.

The "body" of Scorpius (Sco), with red Antares (α Sco), is prominent in the south, with the long chain of bright stars of the "tail" stretching into the Milky Way. In ancient times, the constellation was much larger, but then the "claws" of the scorpion were formed into the independent zodiacal constellation of Libra, to the west.

The main portion of Sagittarius also lies in the Milky Way, east of Antares – in fact the center of the Galaxy is here – but the constellation is ill-defined, except for the central region, which has come to be known as the "Teapot." To the south lies Corona Australis (CrA), not so bright or easy to see as its northern counterpart, Corona Borealis. The much fainter constellation of Telescopium is still farther to the south.

The indistinct constellation of Scutum, with the "Wild Duck" cluster (M11), lies in the heart of the Milky Way between Sagittarius and Aquila. Still farther north, between Aquila and Cygnus, are tiny Sagitta and the larger Vulpecula. To the east lies the distinct, small group of Delphinus.

Equatorial constellations
September, October, November

The southeastern "wing" of Cygnus leads on to Markab (α Pegasi), one of the four stars marking the prominent "Great Square of Pegasus," which is actually a slightly lopsided rectangle, the shorter sides running almost due north and south. (Continuing the line of the "Pointers" in Ursa Major right across the pole actually brings you down the side formed by β and α Pegasi.) However, as the star at the northeastern corner is α Andromedae (Alpheratz), only the other three truly belong to Pegasus (Peg). One other bright star (ε Peg) lies roughly halfway between α Peg and Altair. Between ε and the Milky Way are the tiny constellations of Equuleus, and the much more distinctive Delphinus.

The constellation of Pisces (Psc) is supposed to represent a pair of fish joined by a cord, and lies south and east of Pegasus. The little circlet of the western "fish" is just north of the equator. Much farther north, between Cygnus and Cassiopeia, the small zig-zag constellation of Lacerta crosses the visible boundary of the Milky Way.

The continuation of a line from β Cyg (Albireo) in Cygnus (Cyg) to Altair takes you to α Capricorni, a fairly faint (magnitude 4) visual double, and on to slightly brighter β Cap. Capricornus (Cap) and the next constellation to the east, Aquarius (Aqr), both consist of stars with no very apparent pattern. The only prominent line runs from β Cap through β and α Aqr, with "branches" at roughly right angles to δ Cap and ε Peg. Most of the fainter stars of Capricornus lie south of β and δ Cap. To the east of α Aqr there is a small, distinct "Y"-shaped group of stars, frequently known as the "Water Jar" – from the time when the constellation was regarded as a man carrying water – or more simply as the "Y of Aquarius." To the south and east of α Aqr an irregular line of faint stars completes the constellation.

Farther south is bright Fomalhaut (α Piscis Austrini), which forms the "tail" of a quadrilateral "kite" marked by the stars γ, α and β Gruis, the two latter stars being more or less in line with α Indi to the west. Between this star and the southern part of Capricornus lies the small and very inconspicuous constellation of Microscopium, only two stars of which, γ and ε Mic, are slightly brighter than magnitude 5. Another faint, but rather larger constellation, Sculptor (Scl), lies east of Fomalhaut. Here, only α Scl, outside the chart area to the east, is just brighter than magnitude 4.5.

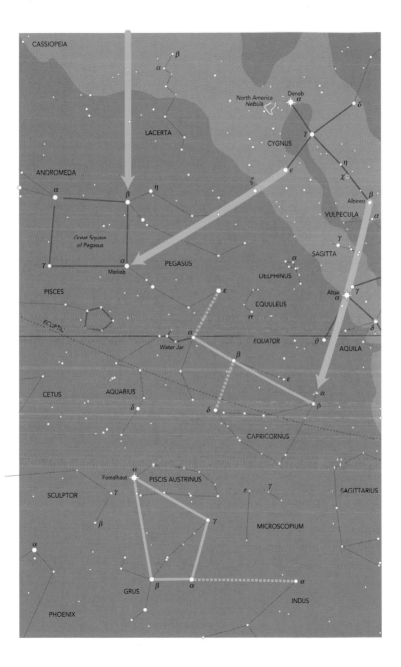

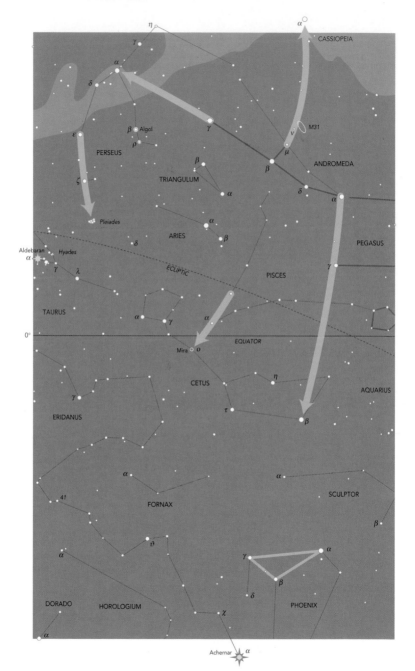

Equatorial constellations
November, December, January

Running northeast from the top of the Square of Pegasus is a prominent line of bright stars formed by the "body" of Andromeda (And), α, β and γ Andromedae, together with α Persei and (with a rather greater gap) α Aur (Capella, outside the chart area). At β And, a side branch of two fainter stars points up toward Cassiopeia. The second star, ν And, lies close to the Andromeda Galaxy, M31, which is distinctly visible to the naked eye as a hazy patch of light.

Perseus (Per) has no easily described shape. One line of stars runs from δ northwestward past α, γ and on to η Per. Halfway between this last star and δ Cas is the famous Double Cluster, h and χ Per: two star clusters that are visible to the naked eye (see chart for northern polar constellations). From α Per a short line runs south to Algol (β Per), a very famous eclipsing variable, and on to ρ, also variable. To the east a line of stars generally trending south begins at δ Per and ends at third-magnitude ζ, above the Pleiades cluster.

Two small constellations lie southeast of β and γ And: Triangulum (Tri) and, with rather brighter stars, Aries (Ari). The eastern "fish" of Pisces (Psc) is just a faint chain of stars running southeast from near δ And to fourth-magnitude α Psc (Alrisha) just north of the equator. Continuing this line south of the equator, we come to Mira (o Ceti), another important variable star which occasionally may become as bright as the third magnitude, although it is more usually about fourth magnitude at maximum. The eastern side of the Square of Pegasus points roughly in the direction of the triangular "head" of Cetus and second-magnitude β. North of the equator, third-magnitude α and a few fourth-magnitude stars form the "tail" of the sea-monster.

In the south the most conspicuous stars are in Phoenix (Phe) and Eridanus (Eri): α, β and γ Phe, and third-magnitude 41, θ and χ Eri, with bright Achernar (α Eri) even closer to the south pole. Between these constellations and Cetus are the undistinguished groups of Sculptor and Fornax.

The constellation of Eridanus represents a river. It begins at a third-magnitude star (β Eri) just north and west of Rigel (see the chart on page 31), and winds its long way south in a chain of faint stars, none of which is brighter than magnitude 3, finally ending at first-magnitude Achernar (α Eri) even farther south than Canopus.

The southern polar constellations

Observers in the southern hemisphere may have no distinct pole star to guide them, but many bright groups and individual stars are circumpolar, or very nearly so, as is a considerable portion of the Milky Way, and the two Magellanic Clouds.

The three first-magnitude and two fainter stars of Crux, often commonly known as the "Southern Cross," are very distinctive. The dark nebula called the "Coalsack" is nearby. Beginners (and northerners) should beware of the larger "False Cross" of second-magnitude stars situated farther along the Milky Way. This consists of the four stars δ and κ Velorum, ι and ε Carinae, rising about three to four hours earlier than Crux itself.

The "upright" of Crux (Cru) points very approximately toward the south pole, but a line from β Cru across to the center of the Small Magellanic Cloud (SMC) comes even closer, passing through the third- and fourth-magnitude group of Musca, close to Crux. This line from the pole to the Small Magellanic Cloud also forms the base of an isosceles triangle, with the central region of the Large Magellanic Cloud (LMC) at its apex.

Rising later than Crux is the brilliant pair α and β Centauri (β Cen being the nearer to Crux), which are unmistakable. Follow a line northwest from β, past second-magnitude ε Cen, to locate ω Centauri, the finest globular cluster in the sky, about 18° from β Cen. From here a line of stars belonging to Centaurus partially encircles Crux on its northern side. The irregular pentagonal shape made by the remaining stars of Centaurus lies still farther north.

Beyond (and south of) α and β Cen is the constellation of Triangulum Australe (TrA), brighter and larger than its northern counterpart. The faint constellation of Circinus (Cir) lies in between, but only α Cir is as bright as magnitude 3. Between Triangulum Australe and the curve of the "tail" of Scorpius the constellation of Ara also consists of third-magnitude stars.

Although many of the constellations in this area have few bright stars, perhaps the most notable figure is the slightly irregular rectangle with one "Alpha" star from different constellations at each corner: α Pavonis, α Indi, α Gruis, and α Tucanae. The fourth-magnitude star β Indi lies between two of the stars on one side, and β Gruis on an extension of the one opposite. Rising later is the triangle formed by α, β and γ Phoenicis.

Tucana (Tuc) itself contains the Small Magellanic Cloud (SMC) and the bright, naked-eye, globular cluster 47 Tuc. The SMC lies on a line between bright Achernar (α Eri) and β Hydri, closer to the pole. Hydrus (Hyi) largely consists of a triangle of third-magnitude stars pointing north, with the apex close to Achernar. Another line of moderately bright stars running north is that formed by γ Hyi, α Reticuli, and α Doradus.

Round beyond the Large Magellanic Cloud (LMC) we return to the stars of Carina (Car), particularly Canopus (α Car) and those belonging to the False Cross, to the south of which lies second-magnitude β Car (Miaplacidus). The shorter "cross-bar" points north to γ Velorum, while two of the stars, δ and κ, are also part of the brightest region of Vela, as mentioned earlier.

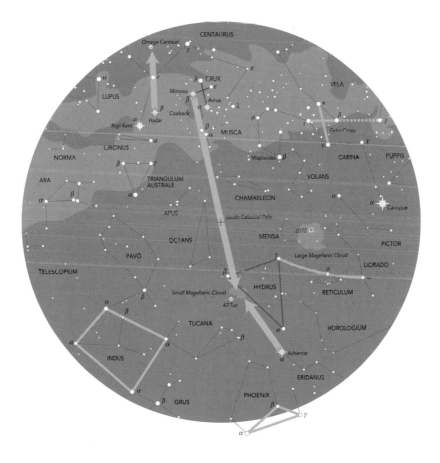

Objects other than stars – what might they be?

It is not uncommon to see something that cannot be identified immediately. Members of the public, unaccustomed to observing, tend to assume that any "light in the sky" that is not the Sun, Moon or a star must be an "Unidentified Flying Object," especially if it moves, flashes or changes color. Astronomers know that there is a surprisingly large number of ways in which even they may be momentarily confused, but that observation, and a little thought, usually enables the cause to be established.

Even when you know a constellation well, it is easy to forget (or even not to realize) that there is a star in a particular position. Variable stars, including rare novae, can sometimes make identification difficult. Planets can certainly confuse, but they are always close to the ecliptic, and usually appear somewhat steadier than stars, even to the naked eye. Aircraft can be very bright if they are using landing lights, and may also seem motionless for a considerable period of time when they are heading toward you. Meteorological balloons can catch the light around dawn and dusk, but usually appear dark against the sky. If watched for a short while they will be seen to move.

Sometimes an individual star seems to move slightly. This may be caused by scintillation, but is most frequently a form of optical illusion which happens to everyone. Many objects do move, of course, the planets slowly, and the Moon somewhat faster. Minor planets and comets may move fairly rapidly if they are close to the Earth, but are usually slow, changing position from night to night. Most minor planets and many comets are also very faint. Balloons, aircraft (which usually show colored and flashing lights) and satellites show faster

What might it be?

Flashing lights	Aircraft, stars and planets with strong scintillation, tumbling satellites
Color changes	Aircraft, stars and planets with strong scintillation
Fleeting, pin-points of light	Head-on meteors (very rare), effects of vision
Steady, unrecognized points of light	Stars, planets, minor planets, variable stars, novae (rare)
Patches of light	Clouds, aurorae, upper-atmosphere experiments, rocket exhaust trails, comets, clusters, galaxies
Visible motion	Birds, aircraft, balloons, satellites, meteors, fireballs
Motion over days (or longer)	Planets, minor planets, comets
Slightly meandering path (satellites)	Effects of vision
Waving changes in position (stars)	Scintillation, effects of vision

► *Fig. 2.4 Strong iridescent colors in the exhaust trail of a rocket launched from Vandenberg Base in California, photographed by Stephen Pitt.*

movement, although with all of these their apparent speed varies greatly with distance and altitude.

Artificial satellites travel slowly in comparison with meteors and fireballs, although their apparent speed does depend on the altitude of their orbit (geostationary satellites are a special case). The slower speed largely distinguishes a satellite re-entry from a fireball. In addition, satellites frequently disappear into (and appear from) the Earth's shadow. Some satellites, most notably the Iridium series (see Chapter 6), produce brilliant, easily visible, flashes of light, which are so distinctive that they are unlikely to be confused with anything else. Finally, it must not be forgotten that there are many night-flying birds – a low, fast-moving bird, dimly illuminated by light from the ground, has been known to appear strikingly like a faint meteor. So, too, has the reflection of car headlights from a nearby telephone wire.

A hazy patch could be one of several things, depending upon its size. The zodiacal light can be seen only along the ecliptic over the eastern or western horizons. Auroral patches, especially at the start of a display, may be taken for clouds illuminated by distant lights. True noctilucent clouds have a very distinct appearance and appear only around midnight during the summer months. Rocket launches and releases of material for upper atmosphere research may give rise to colored glows which could be mistaken for aurorae (or clouds). Normally, however, the rockets' trails are distinctive, and some produce spectacular, brilliantly colored events (Fig. 2.4). In binoculars or telescopes, small hazy patches are usually unresolved clusters and nebulae or galaxies. Such a patch might just be a comet, but this is unlikely, because bright comets are fairly rare.

The usual cause of strong color changes in stars is scintillation, and it occurs especially when they are close to the horizon. Refraction can produce colored fringes on one side of planetary disks, and once again this happens when these are at low altitudes.

The motion of the Moon and planets

Any observer in the equatorial zone is in an excellent position to observe the Moon and planets because at times they may pass directly overhead. However, for observers at other latitudes, their visibility is strongly affected by the season. In summer, when the Sun is high, the region of the zodiac opposite the Sun in the sky must be lowest, and thus badly placed for observation. In winter the opposite is true. The inclinations of the individual orbits to that of the ecliptic mean that each body may be north or south of the path of the Sun. Because of the brightness of the Moon, and its range in elevation of about 10°, its change in altitude is frequently apparent to even the most casual observer. Its continuous motion eastward against the stars (by about its own apparent diameter every hour) is not very obvious to the naked eye, but is easily seen with binoculars or a small telescope. The Moon's movement frequently causes it to hide background stars, and such events, known as occultations, are often extremely interesting to observe, as described later.

When any object is on the opposite side of the sky to the Sun, it is said to be at opposition and is then best placed for observation (Fig. 2.5). This cannot occur with the inner planets Mercury and Venus, and they are easiest to observe at eastern or western elongation, when they appear most distant from the Sun. When, as seen from Earth, a planet appears in line with the Sun (but usually above or below it in the daytime sky), it is said to be at conjunction. All the planets may be on the far side of the Sun from Earth (at superior conjunction), but the two inner planets, Mercury and Venus, may also lie between the Earth and the Sun (at inferior conjunction). It is quite a challenge to observe these planets in daylight around inferior conjunction, when they exhibit narrow crescents (Fig. 2.6). Great care must be taken to ensure safety when

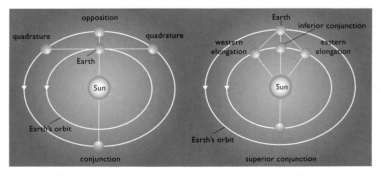

▲ Fig. 2.5 The terms for the positions of the outer (left) and inner (right) planets, relative to the Earth and Sun are shown. Planets are usually invisible at conjunction.

Mercury and Venus are close to the Sun (see page 133). Although they are normally above or below the Sun's disk, very occasionally the inner planets may appear to pass across it in a transit. Again, provided all safety precautions are taken, these events are interesting to observe, and are eagerly awaited by many amateurs. Transits of Mercury and Venus are discussed later, on page 151.

▲ Fig. 2.6 Narrow crescent Venus, photographed by Damian Peach on 2002 November 10, just after inferior conjunction.

The term "conjunction" is commonly used when planets appear close to one another in the sky, or near the Moon or bright stars. Technically, and more correctly, such a close approach is known as an appulse. Occasions when several planets are close together often present a very striking appearance, especially when accompanied by the crescent Moon. They offer an ideal opportunity for photography (Fig. 2.7).

As seen from Earth, all the planets normally exhibit direct motion, slowly shifting eastward against the background stars. However, because of the relative motions and positions of the planets and the Earth, at times they reach stationary points and then reverse direction (Fig. 2.8). Movement westward in the sky is known as retrograde motion. The reversals occur at eastern and western elongations in the case of Mercury and Venus, and on either side of opposition for the remaining planets. Depending on the actual relative positions of the Earth and the planet concerned, the apparent paths may be open ("S"- or "Z"-shaped) or closed loops. Similar effects occur, of course, with minor planets and comets.

► Fig. 2.7 The conjunction of five planets in 2002 May, photographed by H. J. P. Arnold. Jupiter is at top left. Below the cloud at bottom right are (left to right) Saturn, Mars (closest to the cloud), Venus, and Mercury in the twilight arch.

Nearly all the planets rotate on their axes in the same way as the Earth itself (counterclockwise when looking down on the north pole). Only Venus and, strictly speaking, Uranus, are exceptions. All the planets and most of the other bodies in the Solar System have orbits that also follow this same direct rotation. Only among comets and some planetary satellites are retrograde orbits encountered, with the bodies moving in the opposite direction.

Because all the bodies in the Solar System (including the Earth) orbit the Sun in ellipses, their distances from the Sun (their heliocentric distances) vary during the course of their orbits. The point on an orbit closest to the Sun is known as perihelion, and the most distant, aphelion. A planetary disk will naturally appear largest when the distance between the Earth and the planet is least. This effect is particularly important in the case of Mars, where the disk may change

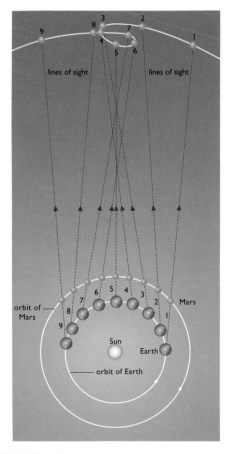

very dramatically in size. The most favorable oppositions occur when Earth is close to aphelion, and Mars is near its perihelion. This can only happen in August or September, when the planet is south of the ecliptic, so southern hemisphere observers are permanently favored in this respect.

The period when a planet is visible is often termed an apparition. It naturally extends on either side of opposition or elongation. Some dates of oppositions and elongations, and approximate positions of the planets for a number of years, are given in tables in the individual sections where planetary observation is described. A table of precise positions (expressed in celestial coordinates) is known as an ephemeris (pl. ephemerides). Such tables are generally given in the astronomical yearbooks for each of the major planets.

◄ Fig. 2.8 As the Earth overtakes an outer planet, the latter appears to reverse relative to the background stars.

—— ASTRONOMICAL EQUIPMENT ——

Binoculars

Binoculars are more useful to beginners than small telescopes, quite apart from being cheaper, and useful for other activities. Their images are the "right way up," and easier to compare with the naked-eye view or star charts than the inverted (or even reversed) fields given by telescopes and the majority of finders (Fig. 3.1). Even the most advanced astronomers frequently use binoculars, and there are some observing programs where other instruments are very rarely employed. The wide field and low magnification make them ideal for observing many star clusters and for general sweeping of the Milky Way, as well as being of use (for example) in tracing the tails of comets when these are too faint for the naked eye, yet too indistinct for larger telescopes and high magnifications.

Choosing binoculars

Old-fashioned opera-glasses, with their very low magnifications, can be helpful on occasions if they happen to be to hand, and are quite useful for scanning the Milky Way. Do not consider buying them, however, if you wish to do serious observing. Prismatic binoculars are far more satisfactory, but several factors (apart from cost) must be taken into account. The most important of these are magnification and aperture (the latter always being given in mm), usually engraved in that order somewhere on the binoculars themselves, as "8× 40" or "7× 50," for example.

▲ Fig. 3.1 Comparison of the size of images obtained by the naked eye and 7× binoculars. The latter are ideal for initial studies because they allow all the major features to be seen. The 6-day-old Moon was photographed by Peter Grego.

Objects for binocular observation

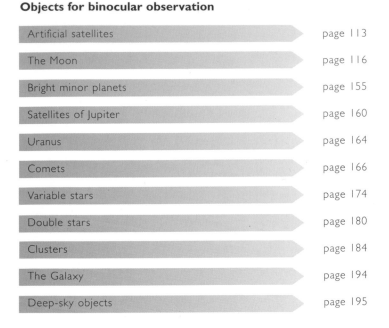

For most general purposes, apertures of 40–50 mm are adequate, although of course, the larger the aperture the fainter the objects that can be seen. Binoculars with apertures larger than 50 mm are certainly desirable for many types of observing, but they are much heavier, and must have some form of mounting to keep them steady.

In the dark, the pupil of the eye is generally about 7–8 mm in diameter. In any instrument, the exit pupil (the diameter of the bundle of rays leaving the eyepiece) should not be much greater than this amount, otherwise light gathered by the objective will not reach the retina. Occasionally binoculars are offered for sale with low magnification and an exit pupil greater than 8 mm. Check this by dividing the aperture by the magnification; 7× 50 binoculars, for example, have an exit pupil slightly greater than 7 mm, which is acceptable. If it is less than about 5 mm, the binoculars have a fairly high magnification, would certainly need a support, and are really more suitable for daytime use. If the magnification is unknown, measure it by the methods given later (on page 66), which also determine the field of view. In most low-magnification binoculars the field of view lies in the range of 5–7°.

High magnifications darken the background and are useful where there is a lot of extraneous light from light pollution, but they simultaneously narrow the field of view, making it more difficult to find objects in the sky. On the other hand, Jupiter's satellites, double stars, and many clusters are easier to see. However, the most important consideration is that binoculars with high magnifications are difficult to hold by hand. Certainly anything over 10×, and more usually over 8×, requires a proper mounting. (Zoom binoculars with their extra optical elements and consequent light-losses are not worth considering for astronomy.) Image-stabilized binoculars are very expensive, but their images are very stable; testing them against ordinary binoculars is an amazing experience

Generally, more expensive binoculars are of sturdier construction, with more rigid mounting of the prisms, making them less liable to misalignment (Fig. 3.2). The more expensive designs use a form of prism, known as a roof prism, which not only provides better optical performance, but also allows a more compact design. The form of the focusing mechanism is another area where a superior (but more costly) design has considerable advantages. Most binoculars use a type of central focusing, with just one eyepiece being adjustable. Individual focusing of the two eyepieces is more satisfactory, because it is more robust. It is, however, much rarer than central focusing. The most expensive designs have an internal focusing mechanism, which provides even better performance. Finally, it should be remembered that in all binoculars there are at least eight glass/air surfaces in each optical train, so full anti-reflection coating is very desirable. It not

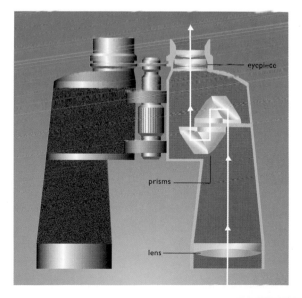

eyepiece

prisms

lens

▶ *Fig. 3.2 The compact design of prismatic binoculars means that they are ideal for many astronomical observations, and can be used anywhere.*

only reduces light-loss, but also improves contrast, and fully coated binoculars will frequently show much fainter stars than partially coated binoculars with the same aperture.

Testing binoculars

Most tests for telescopes (page 59) can be applied to binoculars, particularly those for chromatic aberration, astigmatism, distortion, and flatness of field. Test binoculars by first ensuring that the interocular distance is correctly set for your eyes, and that they are properly focused. You should now see a single image – there should be no sense of the image being doubled. Now gradually move the binoculars away from your eyes to a distance of about 10 cm. The image should remain single, even if you close your eyes for a moment. Choose a featureless object, such a cloudless sky, for the next test. Again, start with the binoculars close to your eyes, but now move them even farther away, to about 30 cm. You will see two distinct disks, one in each eyepiece. These are the exit pupils of the binoculars. They should both appear perfectly round – showing that the full beam of light is passing through the prisms – and evenly illuminated. Any flat sides indicate that some of the beam of light is being cut off. This is a very common fault with cheaper designs.

The alignment of the two optical systems should also be checked, as shown in the diagram (Fig. 3.3). The test is made by again moving the binoculars away from the eyes to about 10 cm, but this time when

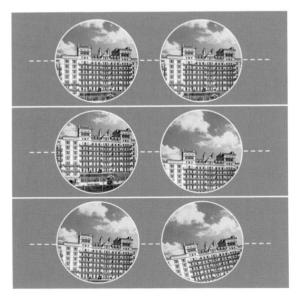

◀ Fig. 3.3 It is important to check that the two optical systems in a pair of binoculars are aligned. With the binoculars held away from the eyes, a straight-edged object should appear perfectly aligned (top). Slight misalignment (center) is acceptable, but angular displacement (bottom) is not.

▶ *Fig. 3.4 Omega Centauri is the finest globular cluster in the sky, easily visible with the naked eye, and a magnificent sight in binoculars or a telescope. It is seen here in a photograph by Chris Cook.*

viewing a distinctive image. Misalignment is a frequent, unrecognized cause of eyestrain and headaches. Your eyes automatically compensate for any slight vertical misalignment, so this is tolerable (although undesirable), but any rotation of one image relative to the other is likely to lead to immediate eyestrain.

Check for anti-reflection coatings by viewing, from both eyepiece and objective ends, the images of a light bulb or strip light reflected by the optical surfaces; colored images indicate coated surfaces, and white reflections are returned by those that are uncoated. Very few binoculars are perfect, so do not be too shocked if you find that your old favorites have some faults.

A test that is often overlooked is to ensure that it is possible to obtain correct focus for both eyes, for an object at infinity. This is particularly important for spectacle wearers, who should carry out a test with spectacles removed. This generally gives better observing conditions than trying to observe while wearing spectacles, although if you suffer from considerable astigmatism, it may be better to continue to use them. In this case, make sure that the eyepieces have rubber eyecups that fold back properly and allow you to see the whole of the field of view.

Ideally, it would be possible to test binoculars at night. The images of stars should appear perfectly sharp and well defined, not only for each individual eye, but also when both are used. This is rarely possible, however, so it is important to carry out tests in daylight, including viewing very distant objects – ones that are truly at infinity.

Observing with binoculars

All telescopes perform better if rigidly mounted, but with binoculars the improvement is startling, much fainter detail and lower magnitudes being seen. Any form of support is better than nothing, so try resting the binoculars on top of a wall, or pressing them against the trunk of a tree. You might like to try a "chainpod" – a length of light chain, fixed to the center of the binoculars and long enough to reach the ground. The binoculars are steadied by putting your foot on the chain and bearing upward slightly. (The same gadget is also extremely useful for steadying a camera for ordinary photography, and you can use a length of strong cord in the same way.) Adaptors can be made or purchased to fit photographic tripods, but these are often difficult to use at high elevations. Many observers find that an ordinary, reclining, garden chair with arms is very convenient and gives good support for the elbows. However, even further improvement can be obtained from a proper observing seat or stand, mounting the binoculars so that they do not have to be hand-held.

Extraneous light can be a problem, but can be solved by shaped, rubber eyecups, which are well worth fitting in any case. Dewing of the objectives can largely be prevented by using dewcaps. The occasional occurrence of dew on the eyepieces is difficult to cure but can be dispersed by fanning the eyepieces with a piece of card.

TABLE 3.1: SOME BINOCULAR OBJECTS		
CVn	M51	"Whirlpool Galaxy" – spiral galaxy (faint)
Ori		"Trapezium" – in Orion Nebula
Peg	M15	Globular cluster
Pup	M46	Open cluster
Sco	M4	Globular cluster
Sco	M6	Open cluster
Sco	M7	Open cluster
Sgr	M8	"Lagoon" – diffuse nebula
Scu	M11	"Wild Duck" – open cluster
Tri	M33	Spiral galaxy
Tau	M45	"Pleiades" – open cluster

Telescopes

In recent years the two traditional types of telescope used by amateurs – the refractor, which employs a lens (or object glass) to form an image (Fig. 3.5), and the reflector, which uses a mirror (Fig. 3.6) – have been joined by various catadioptric forms, which use a combination of one or more lenses and two mirrors (Fig. 3.7). Whatever the type, the main image-forming element is known technically as the objective, and its aperture (D), focal length (F), and focal ratio (F/D) govern the telescope's main applications. Large apertures with their greater light-grasp are usually desirable, because they also give improved resolution and allow higher magnifications to be used. For any particular type, however, large apertures are more expensive and less portable than smaller telescopes.

▶ Fig. 3.5 The standard design of refractor gives focal ratios of f/10 to f/12 while being quite compact (at small apertures), and reasonably portable.

▶ Fig. 3.6 Light in a Newtonian reflector is reflected from a primary mirror to a secondary mirror and directed to a side-mounted eyepiece.

▶ Fig. 3.7 The corrector plate and spherical main mirror in a Schmidt-Cassegrain design produce a long effective focal length in a compact tube.

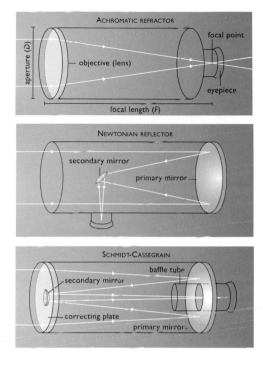

The choice of design is influenced by the type of observing that one wishes to carry out. Gaining experience with binoculars will help you to decide. Although any type of telescope can be pressed into service for any form of observing, some broad guidelines can be laid down. When light-grasp is all-important, reflectors generally score highly; a specific amount of money will usually purchase a reflector with a much larger aperture than a refractor or catadioptric telescope. When resolution is critical, such as in planetary work, refractors are normally superior to reflectors and catadioptrics, aperture for aperture. Catadioptric telescopes score on their ease of use and portability, rather than for any specific optical features. One reason for their popularity, however, is that they perform well on most astronomical objects. A very generalized guide to applications of specific telescope types is given in Table 3.2.

TABLE 3.2: OBJECTS FOR TELESCOPE TYPES	
Reflectors	Deep-sky objects, variable stars, minor planets
Refractors	The Moon and planets, double stars
Catadioptric telescopes	The Moon and planets

Light-grasp and resolution

The most important factor of any telescope or a pair of binoculars is the aperture (D), the diameter of the objective. This controls the light-grasp, which increases as the square of the diameter, dictating the faintest object – the limiting magnitude – that can be seen. A further important quality is resolution, the ability to show fine detail, such as planetary markings or double stars. The resolution is largely dependent upon the size of the objective, and a practical value (in seconds of arc) is given by $138/D$ (the diameter being measured in millimeters). Long focal-ratio objectives (over f/12) may give even better results, closer to $116/D$. Partly because of the presence of the central obstruction, reflectors and catadioptric telescopes usually have slightly lower resolutions than refractors of the same diameter. Reflectors, in particular, can also suffer from tube currents, which may greatly degrade the general performance.

Refractors

The object glass (OG) of a good refractor is achromatic, bringing rays of light of different colors to the same focus. Few amateur refractors operate at focal ratios of less than f/10 or f/12, because achromatic objectives of shorter focal lengths are very expensive. Refractors are therefore best suited to observations requiring long focal ratios, fairly high magnifications, or restricted fields of view. An aperture of 75 mm is about the minimum for "serious" observing. Work can be done with smaller sizes but unfortunately most beginners do not appreciate that the small apertures have considerable limitations, and that good binoculars would probably be far more satisfactory.

The small, cheap refractors often sold through mail order catalogs are generally useless for astronomical work. They frequently have a single lens as an objective (rather than an achromatic doublet), and to mask the dreadful optical aberrations, many have an opaque diaphragm

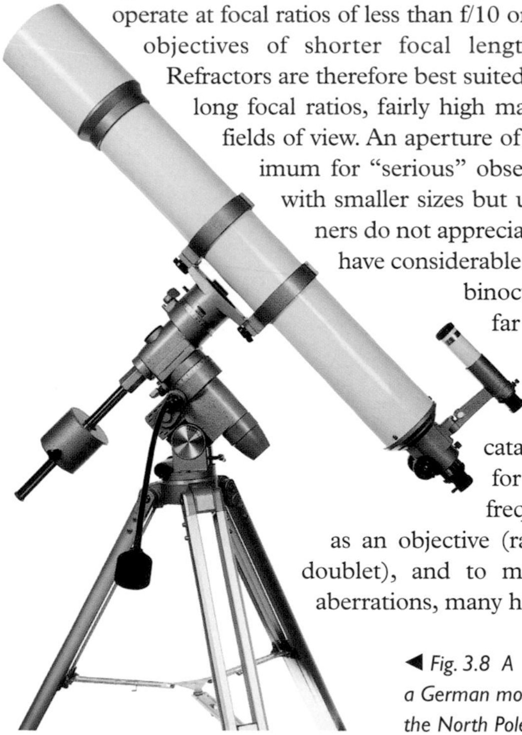

◀ Fig. 3.8 A 100 mm refractor on a German mount, here pointing at the North Pole.

located just behind the objective, with a small central opening. This reduces the effective diameter of the telescope, its light-grasp, and its resolution. Instead of an actual aperture of (say) 50 mm, the true aperture may be no more than about 15 mm. Luckily, such diaphragms are usually easy to see if you examine the telescope from the objective end.

Refractors do not suffer from the central obstructions found in the other types of telescope, which tend to reduce their resolution. Refractors are therefore particularly suitable for the observation of fine detail, such as lunar and planetary features, and for observing double stars. However, a refractor's eyepiece sometimes lies in a very awkward position – particularly when observing overhead – so a diagonal might be required, with the consequent slight loss of light and the problems caused by an image reversal. A refractor of 100 mm is near the limit of portability, and larger sizes are very expensive, so that refractors over 150 mm are rare.

Reflectors

Most amateur reflectors operate at focal ratios of f/6 to f/8, and are generally more suitable for wider fields of view and lower magnifications than refractors. The most usual amateur types are the Newtonian and the Cassegrain. In the Newtonian the secondary mirror is flat, and does not alter the focal length or ratio of the primary. In a Cassegrain the secondary is convex and increases the overall focal length, making it much longer, and thus changes the effective focal ratio of the telescope. Cassegrain types, therefore, have similar uses to refractors and, like them, may require the use of a diagonal at certain times. The wider fields of view available with Newtonians mean that they are ideal for variable-star work – where comparison stars may be a little distance from the variable – and for studying many deep-sky objects or faint minor planets.

Reflectors have the big advantage that, aperture for aperture, they are cheaper than any other type. It is also possible to make the required mirrors oneself, or they can be purchased and mounted in a simple, home-built tube. (The housing for optics in a telescope is always known as a "tube," even though it may have no actual resemblance to one.) When the greatest light-grasp is required, practically all large amateur telescopes (200 mm or more) are reflectors. The smallest useful size for general use is about 150 mm, and this costs about the same as a 75 mm refractor. Such a reflector has a greater light-grasp, showing fainter objects, but it is not as portable as a refractor. Some of the smaller-diameter, short-focal-ratio reflectors usefully bridge the gap between binoculars and ordinary reflectors. They are also easily portable.

One disadvantage of reflectors is that occasionally the reflective coatings need to be renewed, and the optical elements realigned. The two mirrors also need individual dust covers, which must be removed and replaced carefully whenever the telescope is used, unless an expensive optical window seals the tube. The eyepiece of a Newtonian can assume awkward positions, so a rotating tube is a great advantage.

Reflectors are also subject to tube currents – unless they are sealed with an optical window, when they function like refractors – because the air inside the tube mixes with cooler, outside air. Precautions such as making the tube oversize and lining it with insulating material can be effective, but it is for this reason that many reflectors have skeleton "tubes" carrying the optical components. Unfortunately, these may have other problems because they can suffer from air currents caused by warmth from the observer's body – so wear plenty of insulating clothing! – and increased dewing, which may mean that a proper observatory becomes essential.

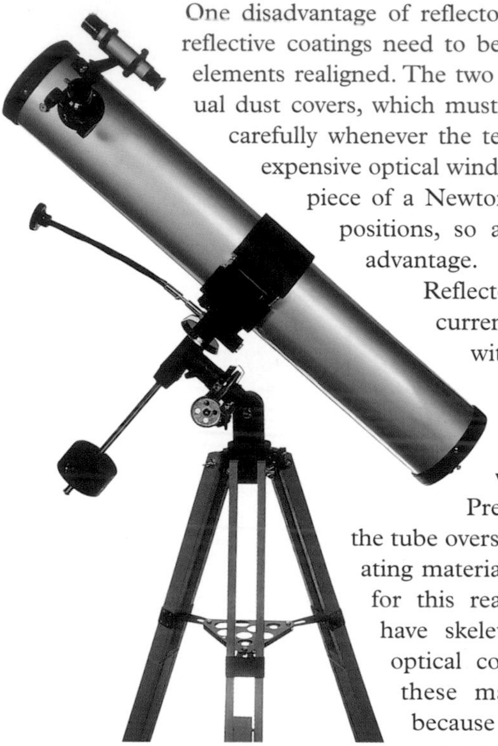

▲ *Fig. 3.9 This reflector's weak mounting may result in vibration.*

Catadioptric telescopes

The Maksutov and the Schmidt-Cassegrain are the most important of the catadioptric telescopes. They are compact for a given focal length and thus very portable and convenient to use, especially as many designs incorporate telescope drives as standard. Despite the fact that they are usually more expensive than refractors or reflectors of the same size, catadioptrics have become extremely popular among amateurs, partly because of the easy portability, but also because of the wide range of features and accessories that have been developed. Their focal ratios are generally long (f/10, f/12 or even f/15), so that they are similar to refractors and Cassegrain reflectors in their applications. Like both of these types, it is sometimes difficult to gain access to the eyepiece without the use of a diagonal.

Testing telescopes and lenses

Certain optical tests can be carried out by anyone, but it must be remembered that no optical equipment is perfect. The errors, or aberrations, must be as small as possible, although the amount that can be tolerated depends upon the type of observations that are undertaken. Planetary and double-star work, and photography, are more demanding than variable-star observation, for example.

To test for chromatic aberration, view a sharp dividing line between light and dark areas, such as the limb of the Moon or Venus. In daytime, view the edge of a distant building, a telephone wire, or the rod of a radio or TV antenna, silhouetted against the sky. There should be no sign of any colored fringes, which would indicate that the different colors are not being brought to a single focus. Reflectors are free from this defect (unless used with very poor eyepieces), but chromatic aberration is often present, even if only to a very small degree, in other forms of telescope, as well as in binoculars.

To test for distortion, scan across any straight line, or examine a rectangular pattern such as a brick wall. Ideally, the lines should remain perfectly straight right across the field and show no sign of any curvature. If lines toward the edge of the field are convex (bulge away from the center), the optics have "barrel distortion," and if concave (curve in toward the center), the cause is "pincushion distortion."

If possible, test any astronomical equipment at night on stellar images, although daytime tests on a distant "artificial star" (such as sunlight reflected from a ball-bearing) will prove useful. Good equipment brings such images to a sharp focus, which under proper conditions appears as a truly circular diffraction disk. The image should remain circular inside and outside focus; any elongation indicates either astigmatism, or possibly that the optical elements are under strain.

Check to see whether a stellar image remains sharp right across the field, without refocusing, or (when examining a star field) if all the images are sharp. The need to adjust the focus between the center and edge of the field indicates curvature of field. Present in most telescopes, it is normally a problem only when

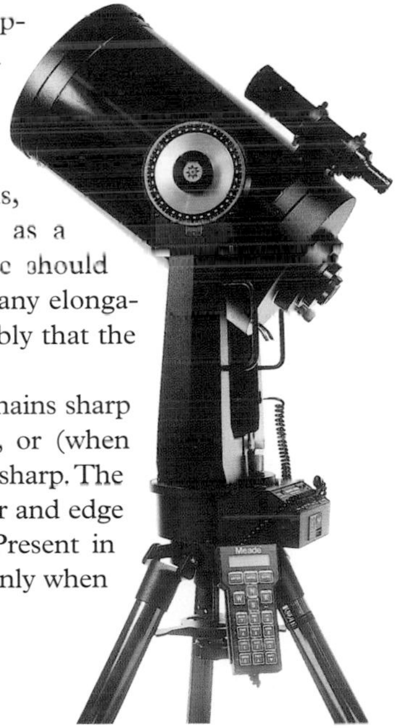

▶ Fig. 3.10 A 250mm catadioptric (Schmidt-Cassegrain) telescope on a computer-controlled altazimuth mount.

they are to be used for photography. Another defect is coma, which is an elongation of images in comet-like shapes at the edge of the field. This is also common, although generally it is more visible in short-focus reflectors than in refractors or catadioptric telescopes with a large focal ratio.

The optical elements in telescopes must be correctly aligned to prevent the aberrations that we have been discussing, and other problems such as vignetting – a loss of light at the edges of the field. All good telescopes are provided with a means of adjusting each of the optical elements individually, in a process known as collimation. The exact means differ from one design to the next, and the task is not one to be undertaken lightly. Beginners are advised to seek advice from their suppliers, the manufacturers, or more experienced friends, before attempting to do this themselves; there are various "tips and tricks" and simple accessory devices that make the task easier and more accurate.

Testing the mechanical aspects of telescopes and mountings largely calls for common sense. Rigidity is essential, both in the actual telescope tube and in the mounting itself, where it is best achieved by sturdy axes each with two well-spaced bearings. Thin, spidery designs, which are prone to vibration, are to be avoided. The rotation on the axes must be smooth, and on equatorial mountings both axes should be provided with clamps. All drives must function without backlash, as should focusing mounts and other movements. Finders and guide telescopes, and the mounting itself, must be capable of fine adjustment (and locking) to permit accurate alignment.

Using telescopes

Any telescope that has been stored indoors will take at least 15–30 minutes to stabilize at the outside temperature. During this time the performance will be poor. Reflectors with thick primary mirrors take much longer to cool down, but if the optics have been well made, the main effect may only be that slight refocusing is required. This causes minimal inconvenience when observing visually, but long-exposure photography should not be attempted until equilibrium temperature has been reached.

Dewcaps – cylindrical or tapering tubes that extend beyond the objective – are simple and save a lot of trouble, but are often forgotten. They are needed on refractors and catadioptric telescopes (and even on some binoculars), and can be made from any suitable insulating material. A dewcap must extend well beyond the objective, like a lens hood, but must not interfere with the edges of the field of view, otherwise vignetting (a loss of light) will occur.

It is most important to try to keep all the optical surfaces clean. A well-fitting dust cover should be used over the objective of a refractor, or the open end of a reflector's tube. In addition, reflectors usually require individual covers over the two mirrors. No covers should be installed until any dew has evaporated completely from the optical surfaces. If possible, a box should be made that will take the whole telescope, but usually this is only feasible for refractors and the catadioptric types. Eyepieces should be removed and stored in a box with a tight lid, giving protection against dust, together with any other small accessories. Although it is often omitted, the focusing mount should also have a cap, to prevent the entry of both dust and spiders. Avoid touching any optical surfaces with the fingers.

If these precautions are taken it should only rarely become necessary to clean the optical surfaces. This might be required perhaps once a year. Remember, however, that scratches are the worst form of damage to any optical surface, and that even visible dust and slight smears will produce little optical effect. Cleaning should only be undertaken when the condition has started to affect optical performance. Frequent cleaning is likely to do far more harm than good.

Treat the optical parts of telescopes, like photographic lenses, with the utmost care; the reflective coatings on mirrors, and the anti-reflection coatings on lenses, are particularly vulnerable. Remove surface dust with a photographic blower, one of the special cans of compressed gas, or by the careful use of a soft-haired (photographic) brush. Never wipe away grit with a cloth or tissue.

If lens surfaces are badly soiled, and they are known to be hard-coated, remove surface dust and then use a photographic, lens-cleaning fluid (preferably with one of the special optical cleaning cloths), lens tissue, or the non-abrasive, photographic, lens-cleaning powder with its special applicator. The eye-lenses of eyepieces are particularly prone to becoming smeared, especially if they are not recessed into the mount. Some eventual deterioration is almost inevitable with such surfaces. Cleaning mirror surfaces (apart from the removal of surface dust) involves washing with a detergent solution, followed by rinsing in distilled water and careful drying. It is best left to the expert. Even with protective overcoatings, reflective surfaces do deteriorate eventually, and will need to be replaced by a specialist firm.

Mountings

There are two main types of mounting for telescopes, of which the first, the altazimuth, can be both simple and cheap. Such a mounting has one horizontal and one vertical axis, which enable it to be moved in altitude and azimuth. In recent years the Dobsonian design of altazimuth mounting has become very popular among amateurs, because it offers a stable, relatively cheap and simply constructed mounting, even with large apertures. However, with all altazimuth telescopes, the orientation of the field of view, and of any finder, alters with the direction in which it is pointed, so location of faint objects can be a problem, especially when charts are being used. Despite this, with experience, quite advanced observing can be undertaken, but photography is generally impossible. (Altazimuth mountings are used for the largest professional telescopes, but they use sophisticated computer systems to control the two axes and simultaneously rotate their detectors at the correct rate to compensate for the changes in orientation.) Ingenious devices have been constructed by some amateurs to allow Dobsonian telescopes to track the stars for short exposures, but these are no substitute for a proper equatorial mounting.

The other form of mounting, the equatorial, has one axis (the polar axis) parallel to that of the Earth, so that the apparent rotation of the stars can be followed without difficulty by a simple rotation at the correct rate. This is essential for proper photography, and enables setting circles and drives to be fitted, quite apart from being generally more convenient to use.

The two main equatorial mountings are the fork mounting and the German mounting. The fork mounting is very rigid, and is normally provided with commercial catadioptric reflectors. It is ideal for Newtonian reflectors, because apart from its other advantages, the height of the eyepiece above the ground is kept to a minimum. Some small catadioptric telescopes are supplied on what might be described as half of a fork mounting – an "L"-shaped support, with the polar axis at the center of the base, and the declination axis carried by the single upright. Although this design allows rapid, and easy, mounting of the telescope, it needs to be extremely sturdy to ensure rigidity.

The other type of mounting, the German mounting, is best suited to refractors or Cassegrain reflectors, because it allows reasonably easy access to the eyepiece at all times. It is a very common, and reasonably satisfactory, design for commercial Newtonian reflectors.

Tripods are frequently used for small telescopes of all types, and are common for small-diameter refractors on both altazimuth and equatorial mounts. They have the great advantage of being portable, and are reasonably successful for fairly large-diameter (150–200 mm) catadioptric telescopes, which, being compact, are reasonably stable. Most manufacturers of Schmidt-Cassegrain or Maksutov telescopes supply suitable tripods for their equipment. If choosing a tripod for a telescope or for astronomical photography with a conventional camera, rigidity is all-important. Generally, the smaller the number of extensions to the legs, and the larger these are in cross-section, the better. If a tripod is regularly used from one location, but has to be removed after each observing session, set location points for each of the legs into the ground. This will enable you to place it in position with the minimum of trouble each time.

In the northern hemisphere it is easy to orientate any portable mounting with sufficient accuracy for all visual, and some photographic work. Locate Polaris in the finder, clamp the telescope in declination, and then rotate it backward and forward in RA. Adjust the alignment until the star remains within about 1° of the center of the finder's field.

In the southern hemisphere the procedure is more difficult, with no bright star close to the pole, but the fainter fifth-magnitude σ Octantis – also within a degree of the pole – can be used in exactly the same way. Long-exposure photographs, however, require more accurate methods of alignment.

With the advent of "GOTO" telescopes, orientation has been simplified. Once the telescope and mounting have been approximately aligned, the telescope is pointed, in turn, at a pair of specific stars held in the drive's database. This provides sufficient information for the drive's computer to compensate for any misalignment, and point

▲ Fig. 3.11 A 150mm reflector on a sturdy mount and tripod (left),

and 200mm and 150mm refractors on a solid pier (right).

the telescope automatically at any other object that the observer subsequently selects. Some designs incorporate GPS (Global Positioning System) components, which determine the telescope's position on Earth and make the appropriate adjustments. Naturally such types of equipment are more expensive than simpler mountings; whether the extra costs are justified will depend on the type of observing being undertaken.

Any telescope benefits from a proper rigid, permanent (or semi-permanent) mounting, and this is certainly required for all large refractors and reflectors, but the performance of "portable" Schmidt-Cassegrain or Maksutov telescopes will also be improved. A metal or concrete pier is very suitable, but must be of an appropriate height for the type of telescope, bearing in mind all the possible elevations that the eyepiece can assume when in use. It is most important that any permanent mounting should allow the equatorial head to be adjusted in both altitude and azimuth, to align the axis exactly with the celestial poles. Too frequently, this is difficult to carry out. If the telescope and mounting have to be removed and stored elsewhere, fix a locating plate to the top of the pier so that the telescope does not have to be aligned every time it is installed.

Telescope drives

All telescopes, including altazimuths, benefit from being provided with a slow motion on each axis, even if only powered by hand. The final position can be adjusted precisely while looking through the main eyepiece. An equatorial mount allows the polar axis to be driven at sidereal rate to counteract the Earth's rotation. Although all forms of mechanisms have been used, nowadays electrical drives are most common, and are found on both fixed and portable telescopes. Electrical safety is essential, and can be ensured by using low voltages (12–24 V). Little power is required to drive even large telescopes if they are well balanced, so vehicle batteries are frequently suitable. If line voltage is used, an isolating transformer or earth-leakage circuit-breaker must be incorporated in the power supply.

Many drive units allow the basic rate of rotation to be varied, and this is particularly useful for guiding during long-exposure photographs. Some also provide lunar or solar rates. Photography is also easier if the declination axis is fitted with a slow-motion drive, enabling corrections to be carried out whilst guiding. Modern telescopes (particularly catadioptric systems) are often available with computer-controlled drives or facilities for finding celestial objects from a built-in database. Such refinements are, of course, fairly expensive, as are auto-guiding systems that will track a given object.

Finders

Most telescopes have restricted fields of view and, even if provided with setting circles (see Chapter 4), require auxiliary wide-field finders to locate objects. A good finder is essential if serious work is to be carried out. The size of the objective and the magnification for such a finder are not critical, although, as with binoculars, a diameter of at least 40–50 mm is very desirable. Magnifications and field sizes similar to those of binoculars will be found to be most convenient, and should be measured in the same way. Try to obtain an image orientation like that of the main telescope. This will make life a lot easier, despite the fact that elbow finders of the type that have roof or pentagonal prisms, and give erect (rather than inverted) images, are very convenient on reflectors. The altered field orientation given by ordinary diagonals, which change with the telescope position, can be highly confusing and most frustrating when searching for faint objects in crowded star fields. "Straight-through" finders, with the option of fitting a diagonal, are best for refractors and catadioptric telescopes, although some of the latter incorporate their own finding systems.

All finders require crosswires, either single or double, to determine the center of the field. Some means of adjustment must be provided so that the finder's alignment with the main telescope can be altered, and locked once correct. Simple focusing arrangements may be required if several observers use the telescope. Dustcaps are advisable, just as with any telescope, and must be completely opaque and well fitting if the main telescope is to be used for any solar work. Dewcaps should also be employed.

Devices – sometimes known as "unity-power finders" – are available that project an image of a graticule into the line of sight, such that an

▶ *Fig. 3.12 A reasonably sized (50 mm) finder on a 360 mm Schmidt-Cassegrain telescope.*

illuminated circle and crosshairs appear to be at infinity, and can thus be used as a finder. This is a useful, modern, and more accurate, version of the old-fashioned, lensless peep sight, which is, indeed, still fitted to some wide-field telescopes that do not require extremely accurate positioning.

Eyepieces

Eyepieces magnify the primary image formed by the objective. There are many different designs, and the type to choose depends upon the focal ratio of the objective, the required field of view, and the type of observation to be carried out. (Cost factors may also be important, because some complex designs are very expensive.) Short-focal-ratio Newtonian reflectors generally require more highly corrected eyepieces than do refractors, catadioptrics or Cassegrain reflectors, if the aberrations are to be kept under control. Most modern eyepieces are of the Plössl type or orthoscopic type, which generally offer good performance and generous eye relief (which is essential for observers who wear spectacles). For wide fields, the Nagler and similar types, which have superseded the earlier types such as the Erfle and König, offer extremely wide fields. (Some even require you to turn your head to see the edges of the field.) The older, less complex (and therefore cheaper) designs such as the Ramsden and the Achromatic Ramsden – frequently, but incorrectly, known as the Kellner (a slightly different type) – are less common nowadays. However, simple eyepieces such as the Ramsden, which do not have cemented lens doublets, can be useful for projecting an image of the Sun, because heat damages cemented eyepieces. As always, fully coated optics are desirable to achieve maximum light transmission and maximum contrast.

Magnification of eyepieces and telescopes

An eyepiece is always specified by its focal length, and you can find the telescope's overall magnification by dividing the focal length of the objective by this number. A 25 mm eyepiece used with a mirror or lens of 1 meter focal length therefore gives a magnification of 40 times. The quoted focal lengths of eyepieces (and telescopes) are frequently incorrect, however, so it is a good idea to measure the magnification directly. First find the diameter of the exit pupil by pointing the telescope at an evenly illuminated surface – the daytime sky will usually do quite well – with an eyepiece in place. Measure the diameter, d, of the illuminated exit pupil as accurately as possible. Divide this into the clear diameter of the telescope's object glass or primary mirror to obtain the magnification ($M = D/d$). This method is easy and provides a reasonably exact value of the combined magnification of the telescope and eyepiece. As mentioned earlier, if the exit pupil is greater than 7–8 mm in diameter, the magnification is too low for the full benefits in light-grasp and resolution to be utilized.

It is just as well to determine the field of view of each eyepiece. An approximation is given by 30° divided by the magnification, but it will vary depending upon the type of eyepiece. Check this at night by first locating a star that is as close as possible to the celestial equator – δ Orionis is a favorite one to use. Point the telescope just preceding the star – that is, west of it – clamp both axes, and record the time it takes for the star to drift across the diameter of the field. Convert this time to degrees and minutes of arc (see the table on page 73) to obtain the field of view. For the wide fields of binoculars or finders this procedure is a little tedious, and it is rarely essential to know the field diameters exactly. In this case, find two stars of known separation, which just fit within the field. These can either be two stars on the equator, or two with essentially the same right ascension, differing only in declination. A cluster with well-determined positions for many bright stars, such as the Pleiades, is ideal for this purpose.

Keep a note of the magnifications and fields of view given by your eyepieces, as this is often helpful, especially when trying to find difficult objects. It is useful, too, to draw the field of your binoculars or telescope finder to scale on a piece of tracing paper or acetate, so that it can be placed upon your charts.

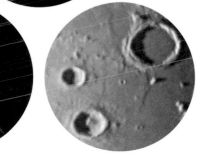

◄ *Fig. 3.13 Typical magnifications used for the Moon. Top: erect 7× binocular image. Bottom: inverted telescope images of 35× (left) and 105×.*

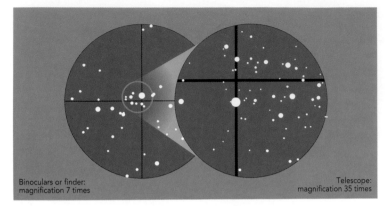

Binocular or finder:
magnification 7 times

Telescope:
magnification 35 times

▲ *Fig. 3.14 A comparison of typical* *magnified, inverted image that,*
magnifications provided by binoculars *depending on the aperture, will*
and a telescope. The latter gives a *show fainter stars.*

Choice of magnification As mentioned under binoculars the minimum usable magnification gives an exit pupil similar to that of the expanded pupil of the eye – 8 mm. Consequently, a 150 mm telescope needs a magnification of about 150/8 (= 18.75) as an absolute minimum. In practice, greater magnification is normally used, except in very specialized applications such as comet or nova searches, which frequently employ large binoculars in any case.

Choose an eyepiece to provide just the required field of view. Nearly all beginners have a tendency to use too high a magnification, but experience shows that resolution is rarely improved, so that higher magnifications do not necessarily show greater detail. In addition, larger images of extended objects such as planets or nebulae are always dimmer, because the same amount of light is spread over a greater area. In theory, the image of a star in a good telescope remains a point, whatever the magnification. In practice this is not always the case, but in variable-star work, for example, a higher magnification may be useful to darken the sky background, or to expand a crowded star field.

A good working figure for the normal magnification is approximately the same as the diameter of the objective in millimeters, with the limit being about twice this amount. The higher magnifications can sometimes be used when seeing conditions are exceptionally good. A useful range of eyepieces for a 150 mm, f/6 reflector, or 75 mm, f/12 refractor (both of which have focal lengths of 900 mm) might be 25 (or 24), 18, 12 and 6 mm, giving magnifications of 36, 50, 75 and 150. Depending upon the exact type of eyepiece, these might have fields of about 50, 36, 24 and 12 minutes of arc, respectively.

Eyepiece accessories

A Barlow lens is a diverging lens that effectively increases the focal length of the objective. It can be useful for increasing the range of a set of eyepieces, or for bringing the prime focus sufficiently far from the telescope for a camera to be used. However, it does not increase the maximum magnification that can be employed, and imposes a penalty in increased light loss, even with full anti-reflection coating. If buying a Barlow, make sure that it does extend the range of all your eyepieces and does not merely duplicate the magnifications that you already have. A Barlow lens with a magnification factor of 1.5× might be more useful than one with a factor of 2×.

An attachment that has the opposite effect to that of a Barlow is a focal reducer (sometimes called a telecompressor). It has become more common since the introduction of catadioptric telescopes, and when used with them gives a lower effective focal ratio, and consequent faster speed for photographic purposes.

Access to the eyepiece of refractors and Schmidt-Cassegrain telescopes can sometimes be very difficult at high elevations. A diagonal, which turns the light path through a right angle, is essential, even though it gives a further light loss, and produces an inversion or reversal, which can be very inconvenient when making drawings, for example. The type incorporating a pentagonal prism avoids this problem, although usually has greater light loss. Not all focusing mounts allow one to be used, because it needs to be much closer to the objective than a normal diagonal or eyepiece.

For planetary and deep-sky work, it is sometimes helpful to use specific colored filters, and a neutral-density filter can occasionally be helpful when observing the Moon, which may otherwise appear blindingly bright under certain conditions of illumination. Many eyepieces are provided with an internal thread to take such filters. Note, however, that it is not safe to use the so-called "solar" filters that are sometimes provided with cheap telescopes and which screw into the eyepieces. At this position, they are close to the objective's focal point, and may fracture with the concentrated heat. It is safer to (literally) throw them away, and use the projection method of observing the Sun.

Observatories

Any permanently mounted telescope requires some protection, and this can be little more than a form of shed which divides, or moves out of the way for observing. However, a proper observatory that gives shelter from the wind will protect the telescope from vibration, and keep the observer warmer. It also helps to prevent interference from nearby lights, and can reduce the problems of dewing. By having everything to hand, and ready to use, more time can be spent observing, rather than transporting equipment backward and forward.

The simpler observatories have roofs that lift off, fold back, or slide away to the side, but undoubtedly a dome (not necessarily hemispherical) is best for giving protection against winds and lights. However, as it has to rotate in azimuth, it is far more complicated, and needs a weatherproof slit that can be opened for observations. All observatories should be opened some time before observing begins, so that that the internal and external temperatures have a chance to equalize. This reduces the local air currents, which can otherwise help to degrade the seeing.

The interior of any observatory needs to be uncluttered to avoid bumping into things in the dark, but it will usually be possible to arrange storage space for equipment, and a flat worktop for charts, handbooks, notebooks, and other material. An observatory clock (showing Universal Time) is essential, as is a permanently mounted, suitably dim red light for illumination when required.

◀ Fig. 3.15 A well-built amateur observatory, with rotating dome, housing a 360mm Schmidt-Cassegrain telescope.

STAR CHARTS

Detailed charts are helpful to any astronomer, and become essential when faint objects must be found. Certain choices have to be made when mapping the sky, particularly the faintest stars to be shown – that is, the chart's limiting magnitude. Because the number of stars increases rapidly toward fainter magnitudes, the number of map sections required to cover the whole sky rises considerably if the charts are not to be too crowded and confusing. Toward fainter magnitudes, too, it becomes difficult to be certain – even in this day of electronic catalogs – that all the stars have been included. So every chart or atlas is a compromise.

Charts and atlases can be prepared showing white stars on a colored or black background (the latter resembling the night sky), or else reversed, with black dots for the stars. Both types have their advantages. The type with a black background (sometimes called a "field" edition) is particularly suitable for use at the telescope when you are seeking very faint objects, because the small amount of white space causes the least loss of dark adaptation of the eyes, which can occur even with a dim red light if large expanses of white paper are examined for any length of time. These charts are also sometimes better for beginners, who find that the representation closer to the actual appearance of the heavens causes least confusion. However, the majority of charts are black on white, and the big advantage is that you can use them to plot any other objects that may interest you. Many atlases are available in both versions. A compromise that works quite well is to have white stars on a blue background, as here.

The charts given here show stars down to magnitude 5, and are suitable for most naked-eye work, where the normal limit is about 6 under good conditions. (About 40 times as many stars are visible

TABLE 1.1 CATALOG DESIGNATIONS		
Greek alphabet and Roman letters A-Q	Bayer letters	Bright stars
Fl 1, Fl 2, ...	Flamsteed numbers	Fainter naked-eye stars
M1, M2, ...	Messier numbers	Brightest clusters, nebulae, and galaxies
NGC 1, NGC 2, ...	NGC numbers	New General Catalogue – clusters, nebulae and galaxies
IC 1, IC 2, ...	IC numbers	Index Catalogue – clusters, nebulae, and galaxies
R, S, etc.; RR, RS, etc.; AB, AC, etc.; BB, ... QZ; V355 ...		Variable stars

in even moderate-sized binoculars.) Binocular and telescopic charts are much more detailed, but normally only show restricted regions of the sky around specific objects. For faint variable stars, for example, a series of such finder charts may gradually lead the observer in toward the variable.

Apart from the Greek, and some Roman, letters given by Bayer, objects on charts and in catalogs are identified by various numbers and letters. Some of the designations most commonly encountered are Flamsteed numbers (usually the fainter naked-eye stars), Messier numbers (clusters, nebulae, and galaxies), and single or double Roman letters, beginning at "R" (variable stars). These, and some others, are listed in Table 4.1.

Celestial coordinates

The way in which the major constellations are identified, and the commonest method of using finder charts, is often called "star hopping" – using patterns of stars to find the objects wanted (Fig. 4.4). There is nothing wrong with this: it is simple and quick, especially for anyone familiar with the sky. But it is also essential to be able to locate any celestial object precisely, and for this purpose a system of celestial coordinates is used. These are right ascension and declination (abbreviated RA and Dec.), which respectively correspond to longitude and latitude on the surface of the Earth (Fig. 4.1).

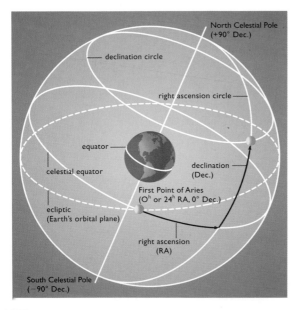

◄ Fig. 4.1 Right ascension (RA) is measured eastward around the celestial equator from the First Point of Aries. Declination (Dec.) is given by the angle north or south of the equator.

TABLE 4.2: CONVERSION OF CELESTIAL COORDINATES TO ANGLES			
RA	units of arc	RA	units of arc
24^h	$360°$	1^m	$15'$
1^h	$15°$	4^s	$1'$
4^m	$1°$	1^s	$15''$

Right ascension is measured eastward along the equator in units of time – hours, minutes, and seconds. The starting point is the meridian (0^h) that passes through the point where the Sun crosses the equator from south to north in March as it travels along the ecliptic. This point, the vernal equinox, is also known as the First Point of Aries ♈, and is as fundamentally important in charting the sky as the Greenwich Meridian is in mapping the Earth.

Declination is measured in degrees and minutes of arc, north (+) or south (−) of the celestial equator. Celestial coordinates therefore range between 0^h and 24^h (= 0^h) in right ascension, and between +90° and −90° in declination.

The coordinates of objects may be determined easily from various charts and are usually given in lists and catalogs. For example:

	RA	Dec.
Sirius	$06^h 45^m$	$-16° 42'$
Andromeda Galaxy	$00^h 43^m$	$+41° 16'$

For most purposes, the stars may be regarded as fixed in RA and Dec. In fact, however, the Earth's rotational axis is slowly swinging round with respect to the stars, mainly as a result of the gravitational effects of the Moon and the Sun. This precession causes the right ascension and declination of the stars to change slightly, but continuously, and explains why the "First Point of Aries," once in that constellation, is now in Pisces. To prevent confusion, charts are therefore drawn for specific, fixed dates, such as the beginnings of the years 1900, 1950 or 2000. These are known as Epochs, and are frequently quoted (in parentheses) after listed coordinates. For example: α Centauri lies at $14^h 39.6^m$, −60° 50′ (2000). The positions quoted here, and the charts, are for Epoch 2000. For most visual work the changes between Epoch 1950 and Epoch 2000 are not very great, and older charts and catalogs may be used fairly easily. However, because precession affects the position of the true celestial poles, the changes must be taken into account when aligning telescope mountings for long-exposure photography.

The RA on the meridian for any observer at any particular time is identical to the Local Sidereal Time (page 85). It is often convenient

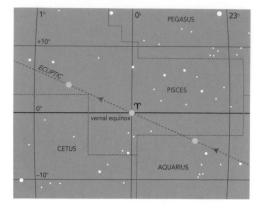

◄ Fig.4.2 The vernal equinox (First Point of Aries) is now in Pisces and slowly moving along the ecliptic toward Aquarius.

to know the hour angle (HA) of an object, that is, the difference between its RA and the RA currently on the meridian. Strictly speaking, hour angle is measured westward (in the same units as RA), because it increases with time, and it may be necessary to add 24^h to obtain the correct value. However, it is frequently regarded as applying in either direction, and given as hour angle west, or hour angle east, of the meridian.

▲ Fig. 4.3 The vernal equinox is currently slightly east of the circlet of stars that form the western "fish" in the constellation of Pisces.

STAR CHARTS

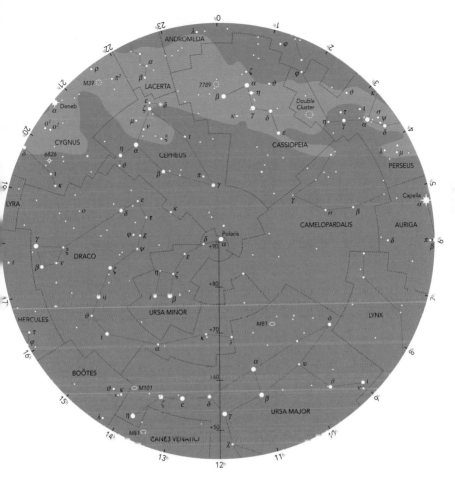

Key to star charts

	Magnitudes								Ecliptic
	−1	0	1	2	3	4	5		
Double stars			Diffuse nebulae						Constellation boundaries
Variable stars			Planetary nebula						
Open star cluster			Galaxies						Constellation figures
Globular star cluster			Milky Way						

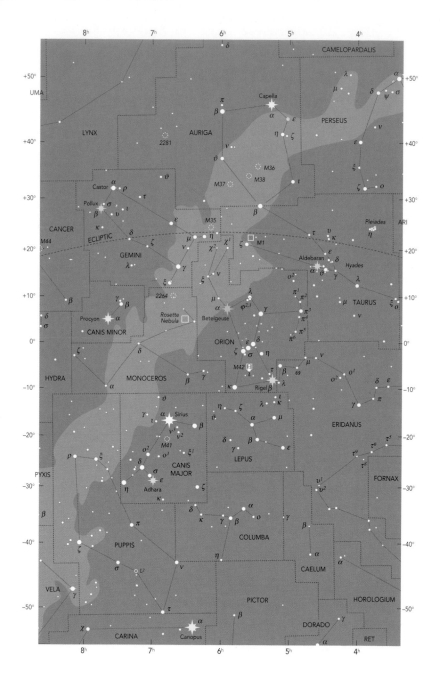

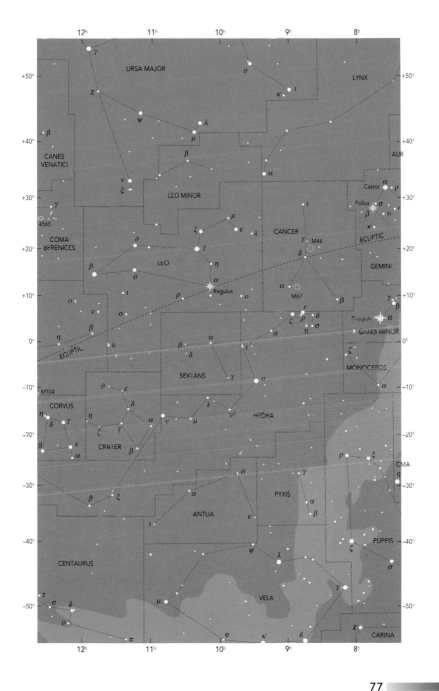

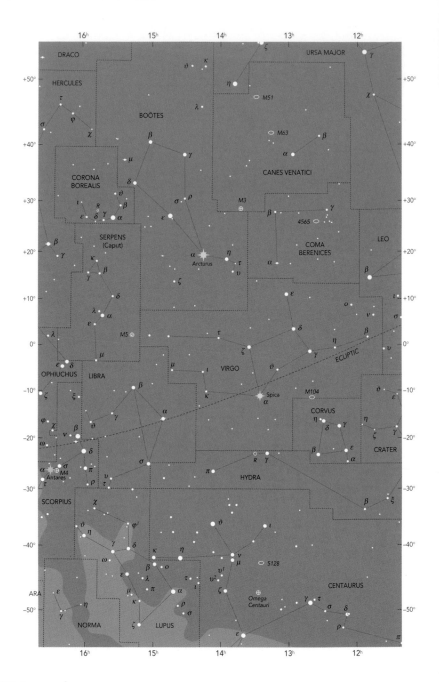

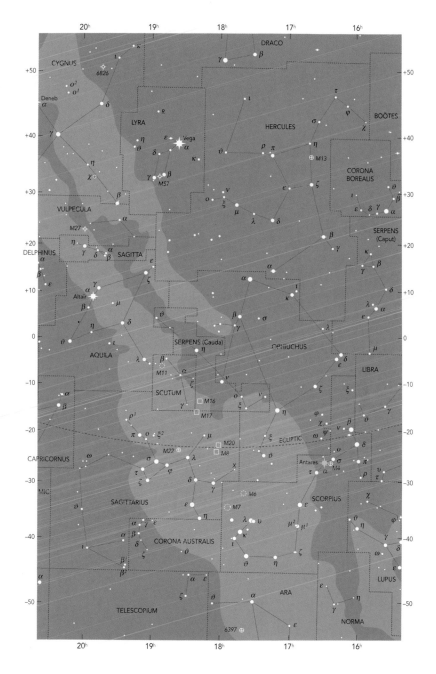

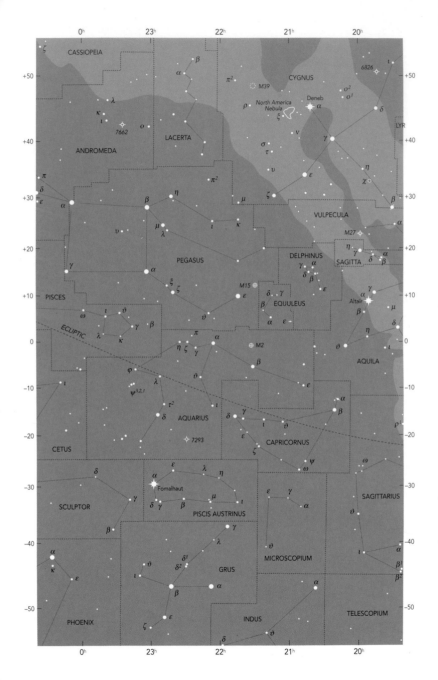

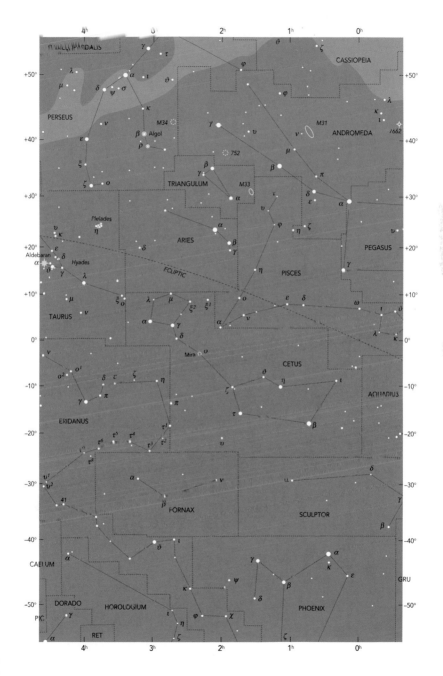

Finding objects with binoculars and telescopes

Comparing charts with the sky can sometimes be very confusing. Binoculars or a telescope usually show more stars than are marked, the apparent scales are different, and the orientation may not be obvious. With regional charts (such as those in this book) turn them to match the sky. Depending upon the equipment with which an object is likely to be observed, finder charts may be prepared with north at the top (for binoculars), or south at top (for telescopes). An inverted field rarely causes great problems, but diagonals may give strange reflected or inverted images so that charts have to be viewed from the back against a light. Avoid diagonals, especially on finders, until you are familiar with the field sizes of your telescope and finder.

Looking through an eyepiece, it is generally possible to recognize a pattern of brighter stars somewhere in the vicinity of the object, and charts can be turned if necessary, so that their pattern matches the sky. If you have great difficulty with a telescopic field, perhaps in the crowded regions of the Milky Way, it frequently helps to identify guide stars first with binoculars, then pick up the same stars in the finder. Making a drawing of the field can also help with positive identification.

If you have no chart with which to "star hop," you need to know the coordinates of any object you want to find. Setting circles – described below – make the job easy, but there are methods that can be used without them. Sometimes positions are given relative to bright stars, using the terms preceding and following. The former indicates that the object precedes the star across the sky. In other words its RA is less, and it lies to the west. In an inverting telescope it will appear toward the left of the field. The opposite is, of course, true for "following." On other occasions you may need to calculate the offsets from listed positions. Remember to convert the difference in RA into degrees (Table 4.2,

► Fig. 4.4 In "star hopping" (top), the patterns shown by the brighter stars are used as a guide to the position of fainter stars and eventually to the object required. When the field size is known it may be used (center) to locate faint objects from brighter, easily visible ones. Bottom: Other useful methods of finding objects are sweeping in RA (A) or Dec. (B), and offsetting in both coordinates from stars of known position (C). With an inverted field, the galaxy is following (east of) star A, and north preceding star C.

page 73), then step across from a bright star using the field diameter (page 67) as a guide. This may be done with any equipment, although it is far easier with an equatorial, when true "sweeping" can be used. Look up a bright object with the same RA or Dec. as the object you want. If the RA is the same, center the bright star, clamp the RA axis, and sweep north or south in declination. Clamp the other axis if declination is the known coordinate, and sweep in RA. If you still have trouble, there is yet another trick you can try. Find something with the same Dec., but preceding the object. Clamp both axes and wait the appropriate amount of time until the Earth's rotation brings the elusive object into the field of view. (You can, of course, do this with an altazimuth mounting, but only on the meridian.)

Setting circles

If an equatorial mounting has been properly aligned with the celestial pole, graduated setting circles can be used to point the instrument at any object. These circles should be as large and accurate as possible. A declination circle 150 mm (6 inches) in diameter might have degrees and probably 30′ subdivisions, and a similar right ascension circle would have hours, subdivided into 2-minute intervals. Each circle is provided with an index against which a reading is taken. Declination presents little problem, because the circle may be read directly, or else used to offset from an object of known declination. Similar offsetting is possible in the other coordinate by simply taking the difference in RA and essentially using the circle as a protractor.

Finding the right ascension directly is slightly more complicated because it involves knowing (or finding) the sidereal time (page 85), and depends upon whether the RA circle is fixed to the polar axis, or is adjustable. Taking the fixed case first, the circle must read 0^h with the telescope pointing due south. Find the hour angle (HA) of the desired object, and turn the telescope the appropriate amount east or west of the meridian. When the RA circle is adjustable, center a bright star in the telescope and turn the circle until it reads the (known) RA, then clamp it to the axis. Now turn the telescope until the index reads the RA of the wanted object. This arrangement involves adjusting the RA circle for every new object, but is very suitable for small telescopes. Some telescope drives turn the RA circle to follow the stars, so that, once set, the telescope's right ascension may be read at any time. Even more sophisticated (and expensive) drives have a built-in database of objects, so that finding your target simply becomes a matter of entering its name and letting the telescope locate it for you. Such "GOTO" drives are more commonly found on catadioptric telescopes than on conventional reflectors and refractors.

——— RECORDING OBSERVATIONS ———

Time

The time used for ordinary civil purposes is a Mean Time (MT) based upon the average length of a solar day. The actual day length varies throughout the year because of two factors: first because the Earth's speed in its orbit is not constant, and second because although the Sun appears to move at a fairly regular rate along the ecliptic, the length of day is affected by its changes in right ascension, which differ at different times of the year. The amount that must be added to mean time to give apparent time (as would be shown by the Sun on an ordinary, simple sundial) is known as the Equation of Time, or E.

Time based upon the "fictitious" mean Sun gives a Local Mean Time (LMT) that varies with longitude around the world. However, ordinary clocks indicate standard time based upon various standard time zones, although the use of Daylight Saving Time (Summer Time) may complicate the issue. This occurs in North America, where Hawai'i, most of Arizona and Saskatchewan, and parts of British Columbia, Nanavut, Ontario, and Quebec do not employ Daylight Saving Time.

The time used for reporting astronomical observations and used in handbooks and almanacs is Universal Time (UT), otherwise known as Greenwich Mean Time (GMT), the standard time at the zero (or Greenwich) meridian. It is reckoned from 0^h to 24^h, beginning at midnight. UT always uses the 24-hour clock, avoiding problems that occur with the use of "a.m." and "p.m." Because UT is the same for everyone, everywhere, it avoids a lot of confusion. Every good observatory (and observer) has a clock showing this time. You may obtain UT from your local (winter) standard time by using the time-zone diagram shown in Figure 5.1.

If you observe any unusual phenomenon (such as a daylight fireball, for instance) outside a normal observing session, it is usually best to record the time as Local Standard Time (or local Summer/Daylight Saving Time), and make the appropriate correction to Universal Time later. In the excitement of the moment it is only too easy to make a mistake in adding or subtracting the correction.

Another system of time, Sidereal Time (ST) is based on the interval between two successive transits of a star across the same meridian. A sidereal day is approximately 3 minutes and 56 seconds shorter than a mean solar day. Like ordinary mean time, sidereal time is reckoned from 0^h to 24^h, but it begins when the vernal equinox, "the First Point of Aries" (page 73), crosses the meridian. The Local Sidereal Time (LST) is given by the right ascension on the meridian. This can be established by observing the transit of a star of known right ascension across the meridian. You can then set an ordinary clock to that time, which will be sufficiently accurate for a single observing session, and can be used for finding objects with setting circles. However, it is very useful to know your local sidereal time at any particular moment, so see if you can obtain a clock that will run fast by about four minutes per day. Some old-fashioned (that is, mechanical) clocks can be adjusted by this amount – most battery clocks cannot. Many handbooks give Greenwich Sidereal Time (GST) for 0^h UT. Your longitude (converted to hours, minutes, and seconds – see Table 4.2, page 73) gives the difference between your own, individual local mean time and UT, and also – to a sufficient degree of accuracy – between your local sidereal time and Greenwich Sidereal Time. Most of the planetarium programs available for use on personal computers can be set to show Universal or Local Time (and often Sidereal Time), although regrettably, one or two programs persist in showing (and requiring one to enter) times as a.m. and p.m.

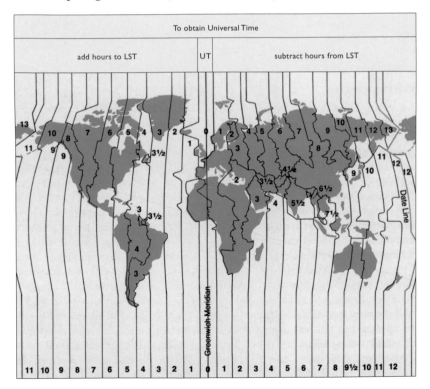

▲ *Fig. 5.1 The standard time zones of the world may be used to obtain* *Universal Time (UT) from the observer's local (winter) standard time.*

Dates

To prevent confusion, there is an internationally accepted, scientific way of expressing dates, where the elements are given in the descending order: year, month (in letters, not figures), day, hour, minute, and second (UT). So you might see, for example, "Second contact of solar eclipse – 2012 November (or Nov.) 13, 20:37 UT" (no seconds in this case). Occasionally you may also come across time expressed in decimal days, that is, in decimals of a day, taken to the required degree of accuracy. (The decimal equivalent of the time just quoted would be 2012 Nov. 13.834028.) This form is particularly helpful for carrying out calculations.

It is very convenient to be able to keep track of events that occur a long time apart, such as the appearances of a comet, or the maxima of variable stars, to give just two examples. The ordinary civil calendar, with its unequal months and leap years, is not very suitable. So astronomers often use Julian Days (JD) for their records. In particular, most variable-star observers use JD all the time in reporting observations because it makes the production of light-curves so much easier. In this method of calculating the date, the days are numbered from 4713 BC January 1 – a time so far in the past that no earlier observations will ever be available. The days start at 12:00 hours (noon) UT, not midnight. Julian Day 2,455,928.0 began on 2012 January 1, 12:00 UT, for example. So the observation of the total solar eclipse, previously mentioned, on 2012 November 13, 20:37 UT would be on JD 2,456,245.359028. Tables of Julian Day numbers often give the date for day zero in each month, an apparently strange idea, but which actually makes life easier, because for observations after 12:00 UT you can just add the date. Frequently, you only need to use the last few figures of the JD, especially when there is little chance of confusion.

Making detailed observations

It is important to keep a record of your observations – even when you are just starting. Never forget that you could be the only person observing a particular object at a certain time, and that even a rough sketch or just a few details might give information that no one else can provide. This is one of the reasons why every astronomer is encouraged to submit observations to one of the national or international amateur organizations (pages 213–214). Beginners often feel that their work could not possibly be "good enough." This is not true. It is not difficult to make good observations; it merely requires practice for the procedure to become simple and quick. This applies to both "numerical" observations, such as timing events or making magnitude estimates, and to "artistic" ones, like planetary drawings. When beginners' observations are compared with those of experienced workers they are often found to agree exactly. Someone's very first variable-star estimate, for example, has been precisely the same as one made by an observer with many years' experience – the only difference was that the former took much longer to make the observation, and was not so confident of the result.

Keeping records

Have one observing logbook (or possibly one for each main type of object) with fixed pages into which details are entered at the time. (Loose-leaf folders are useful for organizing secondary copies of the main observations.) For each observing session include the date (civil or JD), the time of individual observations, seeing conditions, and equipment used (telescope, magnification, and so on.). Follow the internationally accepted method for recording dates and times, giving them in the descending order: year, month (in letters), day, and time (UT). (If for any reason you have to use a local or summer time, this must be stated in the book.) If the date may change during an observing session, quote the "Double Date" at the beginning (e.g. "1985 Aug. 12/13") to prevent confusion.

The actual details recorded will of course depend upon the objects observed, and must be entered at the time. Try to avoid being biased by your own (or other people's) earlier observations. This is not always easy: if you saw a certain pattern of sunspots yesterday, for example, you tend to expect them to be the same when you look today. So record only what you see, not what you think should be there. Never alter details later if they seem wrong, or appear to disagree with anyone else's work. Even very experienced observers make mistakes, and there are always differences between individuals. "Doctored" observations are worse than useless. If you do spot an error, make a note in your logbook (and on any other copies), so that neither you, nor anyone else will be misled.

Some experienced observers have developed the habit of writing down their observations in complete darkness, to avoid any possibility of losing their dark adaptation. They do this by using a soft pencil on a plain notepad and drawing a thick black line under each observation. Even though the actual writing may not be clearly visible, it is surprisingly easy to write legibly in complete darkness. The black lines under each observation are sufficiently distinct to be detectable in faint starlight. If this method is adopted, however, the observations must be transcribed exactly (errors and all) into the permanent observing notebook.

There is another point worth mentioning – take your time. Do not rush just because another observer makes several observations while you struggle with one. Your one observation may be worth more than all the others. (Of course, some work has to be done quickly, but that is a different matter.)

How to make drawings

If in doubt, make a drawing. That is good advice for any observer, even those who are not looking at planets or similar objects. If a galaxy, for example, proves to be invisible, make a little sketch in the observing book of the surrounding field. This can show that you were (or were not!) looking in the right place, and helps with identification at a later date. Planetary satellites, minor planets, and similar subjects may be treated in the same way. In any case, the concentration required to produce a drawing forces the observer to pay more attention to the object, and usually results in more detail being seen.

Detailed drawing requires care and patience. It is best to start with something fairly simple, without too much detail – as mentioned before, drawing the naked-eye view of the Moon is quite good initial practice. Planetary disks do not show many features in small telescopes, and are probably easier to draw than a highly detailed telescopic image of the surface of the Moon. Most whole-disk drawings are made to standard diameters (given later), but other subjects such as portions of the Moon should never be made too large: sizes of 100 to 150 mm (4 to 6 inches) square as a maximum. On the other hand, do not cramp yourself for room, and do not attempt to draw too much at once.

Not much is needed in the way of equipment: a clipboard perhaps, some good-quality drawing paper, soft pencils (generally 2B and 4B), erasers – the pencil type is useful – and some "stumps" (small pieces of blotting paper rolled into narrow cones). Colored pencils are excellent for representing Mars, Jupiter and Saturn, and of course

▶ Fig. 5.2 An ideal way to start is with black and-white drawings of lunar craters, such as this one of Pythagoras by Robin Scagell.

▶ Fig. 5.3 Although color (as in this drawing of Venus by David Graham) is not essential, it may be added as you gain experience.

there are many other varieties of medium that may be tried. If ink is to be used, as in lunar work, a denser board may be required, rather than normal drawing paper.

There is no reason why the initial sketches should not be quite rough and carry notes about the position, shape or intensity of the various features. A "clean" version may then be prepared later – even away from the telescope. Lay down the broad outlines of the features first, and gradually refine them. The stumps – or your fingers! – can be used to spread out the pencil to give the correct shadings. Use the pointed eraser to pick out small lighter areas. With the Moon, outline drawings of craters are easier for the beginner than trying to reproduce the exact appearance of the highly contrasting features (Fig. 5.4).

A certain amount of work can be carried out away from the telescope, such as using Indian ink to deepen lunar shadows, or Chinese White to pick out brilliant highlights. Ideally, however, the finished drawing should be checked against the actual appearance through the telescope. The dark sky background can also be added, usually in black, although blue is sometimes employed for daylight observations of Venus, for example. Again, if ink or color washes are to be used, a heavier board is needed rather than just drawing paper. All pencil drawings should be sprayed with fixative to prevent their being smudged.

If only a few drawings are ever made, glue them into the observing notebook in the appropriate places. (It is best to fix only one edge rather than attempting to paste down the whole of the back, which usually only results in wrinkles.) Make sure, however, that all drawings, and most especially those of planets and comets, carry full details, just in case they do ever become detached. If planetary observation becomes your main interest it may be a good idea to choose a notebook

▲ Fig. 5.4 Stages in drawing the craters Steinheil and Watt (by Peter Grego). Basic outlines are sketched in first, taking care to position the features accurately. Shadows are then added, followed by fine detail and tone.

(or notebooks for each planet) with fixed, blank leaves of drawing paper. Suitable planetary outlines may be prepared just before starting to observe. Planetary observation is discussed in greater detail later.

How to take astronomical photographs

There is a fascination in trying to obtain good photographs of celestial objects. Once again, it is not essential for you to have expensive or complicated equipment to produce satisfactory results. Your camera must, however, be capable of making time exposures. Some modern, single-lens reflex cameras cannot be used because they rely on battery power to raise the mirror, or keep the shutter open. The batteries can go completely flat in the middle of a single, long, astronomical exposure. Some types of observational work use quite ordinary, unsophisticated cameras, without any form of special mounting. On the other hand, some amateurs find that faint objects such as distant galaxies pose a great challenge, requiring large telescopes, special films and equipment, and non-standard processing techniques.

In recent years, amateur astronomical imagery has been revolutionized by the introduction of CCD (charge-coupled device) cameras, which allow amateurs with good equipment to obtain images, and reach limiting magnitudes, that just a few years ago were only attainable with professional instrumentation. Not only have amateurs obtained some extremely beautiful and striking images, but they have also scored considerable success in discovering supernovae by this method, as well as detecting faint transient objects, such as those known as gamma-ray bursters, which were once exclusively the preserve of professional astronomers with large telescopes.

Given that CCD imagery requires moderately expensive equipment, computer processing, and specialized techniques, it is not discussed here, although some of the images reproduced in this book have been obtained in this manner. Digital cameras can also be used for astronomical imagery, but again, discussion of the techniques is beyond the scope of this book. A summary of the types of objects that may be photographed with different methods is given in Table 5.1.

TABLE 5.1: PHOTOGRAPHY	
Fixed camera	Star trails, constellations, meteors, aurorae, noctilucent clouds, artificial satellites, sequences of Moon or lunar eclipses
Driven camera (unguided)	Moon, lunar eclipses, constellations, clusters
Driven camera (guided)	Star fields, nebulae, comets (wide-field), minor planets
Telescope + camera & lens	Lunar features
Telescope at prime focus	Star fields, nebulae, clusters, comets, galaxies
Eyepiece projection	Planets

Undriven cameras

It is easy to begin by photographing star fields with any ordinary camera that allows time exposures, on a fixed mount, such as a photographic tripod. This produces trailed images of the stars, whose length naturally depends upon the exposure, and the distance of the area concerned from the celestial pole, where motion is least (Fig. 5.5). The standard, 50 mm lenses in most 35 mm cameras give fields of about 39 × 26 degrees, which is sufficient to include most individual constellations. With moderately fast color or black-and-white films (say about 200 ISO), even short exposures of 20 to 30 seconds will show about as many stars as can be seen by the naked eye, if not more, and the trails will be short. Stars are a very severe test of any optics, and even good-quality lenses may show strangely shaped images at the edges of the field when used at full aperture. It may be best to use a smaller aperture, and as with all other astronomical photography, this can only be settled by experiment. Similarly, the length of exposure will be partly determined by the brightness of the sky background, whether this is due to artificial lights or moonlight. Keeping a note of the exposure details, together with the usual information about date and time, will help you decide what is best for your conditions.

Camera-shake is a problem in any astronomical photography, so a cable or pneumatic release must be used. The motion of the mirror in single-lens reflex cameras can cause vibration. Some designs allow the mirror to be locked up, and some immediately raise the mirror if the self-timer is used, so the vibrations die away before the shutter is released. If neither feature is available, hold a black card, a black hat or something similar in front of the lens as a simple shutter.

◀ Fig. 5.5 Long star trails produced by a time-exposure of the eastern sky in the northern hemisphere.

▶ *Fig. 5.6 A camera and telephoto lens mounted piggy-back on a large refractor. The latter's drive and guiding facilities may be used during long exposures.*

If it is difficult to see faint stars through your camera's viewfinder, make a simple wire-frame to outline the field of view for the standard lens (or lenses). Fix it to the camera, or its mounting plate, and locate the position for your eye by checking the field in daylight. You can use a fixed peep-sight to mark this distance for use in the dark. This device can also be used for driven cameras.

Driven cameras

To avoid trails, some form of equatorial mounting is needed, so a camera is frequently mounted on a driven telescope (Fig. 5.6). If this is not available, a simple mounting carrying just a camera is easy to make. The stars can be tracked quite successfully with only a hand drive, so the mount can be fully portable and used anywhere, but like telescope mountings, electrical drives are frequently used. However, any mount must be aligned reasonably well with the pole. The longer the focal length of the lens used, and the longer the intended exposure, the more accurately this must be carried out.

Even with a good-quality drive, it is usually necessary for corrections to be made during a long exposure, especially when long-focal-length lenses are used. Various errors accumulate and cause the image to wander from the correct position on the film. When the camera is mounted on a large telescope, the latter can be used for guiding. A mounting designed just for cameras can also carry a small, long-focus refractor for this purpose. In both cases guiding can be on any bright star, because the camera and telescope can point to slightly different regions of the sky.

Photography through a telescope

Any astronomical photography is greatly affected by the seeing conditions, but these become particularly important when telescopes and very long exposures are used. The errors caused by atmospheric effects are usually far smaller than those produced by incorrect polar align-ment, mechanical problems, and faulty guiding (amongst others).

You can obtain photographs by mounting a camera (with its lens) in line with a telescope's eyepiece – both being focused to infinity. However, better results are produced if the light from the telescope is focused directly on to the film in just a camera body or in a specially made film-holder. This prime-focus photography makes the best use of the available light, but the image scale is governed by the focal length of the telescope, and is approximately given (in mm per degree) by the focal length divided by 57.3. A telescope of 1200 mm focal length (150 mm, f/8 for example) has a prime focus scale of about 21 mm per degree, and a full 35 mm camera frame would cover about 1.7×1.1 degrees. This is much greater than any normal, visual field, and field curvature is certain to be evident, with out-of-focus images at the edges of the frame. (Photographic lenses are specifically designed to have flat fields.) Other aberrations, especially coma in the case of Newtonian reflectors, are likely to be present at the edges of the field. The field becomes flatter, and the other aberrations are often reduced, as the focal length increases, so changing the effective focal ratio can help.

The focal length can be increased by using a Barlow lens, or by using eyepiece projection, which will also produce a larger image. This is frequently essential in any case – especially with planetary photography. The size of the image of the Moon, which has an angular diameter of about 31 minutes of arc, is about 10.9 mm at the prime focus of a 1200 mm telescope. (The Sun is about the same, but solar photography

▲ Fig. 5.7 A photograph by Akira Fujii of the Large and Small Magellanic Clouds. The second-brightest globular cluster, 47 Tucanae, lies near the Small Cloud (left).

requires special precautions – see page 133.) The planets with the largest apparent sizes, Venus and Jupiter, can only reach about 60 and 45 seconds of arc respectively. Their sizes would only be about 0.35 and 0.26 mm in diameter, far too small to be enlarged in the darkroom. But any magnification of the image means that it becomes fainter, requiring a longer exposure with all that this entails.

▲ *Fig. 5.8 A video image of Jupiter obtained by Steve Massey.*

With a reasonably large-aperture telescope having a long focal ratio, the use of a focal reducer (page 69) can be of considerable help in obtaining photographs of extended objects such as nebulae (Fig. 5.9). The shorter effective focal length and faster focal ratio allow shorter exposures – but the images will be correspondingly smaller, and require more enlargement.

Adaptors are available to couple most interchangeable-lens camera bodies to the drawtubes of focusing mounts. One of the greatest problems with photography through a telescope is obtaining the correct focus. Most focusing screens are unsuitable, because they are designed for use when there is plenty of light, although they can be satisfactory with very bright objects, such as the Moon. If you have one of the more expensive cameras with interchangeable screens, you may be able to use the type with a clear center and engraved crosshairs. With an eyepiece magnifier in place, first make sure that the image of the crosshairs is sharp. This can even be done off the telescope, by pointing the camera, without its lens, toward an evenly illuminated surface – a piece of paper, for example. Do not change the focus of the magnifier afterward. Looking through the magnifier, adjust the telescope's focusing mount until both the image and the crosshairs appear sharp.

With other cameras it may be necessary to focus at the actual film plane. A focusing plate like the screen mentioned (clear center and crosshairs) is ideal. It can be made at home by drawing thin Indian ink lines on a finely ground piece of glass. Hold the plate on the film plane and focus in the way just described, preferably with a magnifier. Another method is to use a "knife edge," again at the film plane. With the telescope pointing at a bright star, and the eye held back from the camera, an illuminated circle is visible. With the motion of the star across the knife edge – let the rotation of the sky do this – the image will darken evenly all over, rather than from one side, only when the knife edge is at the precise point of focus. (This is actually a mirror-testing technique known as the "Foucault Test.") Obviously both of these methods are more suited to old-fashioned plate cameras than the modern 35 mm

◀ Fig. 5.9 Two emission nebulae in Sagittarius: the Trifid Nebula, M20 (top), with a small, blue reflection nebula, and M8 (bottom), the much larger Lagoon Nebula, photographed by Stephen Pitt.

variety, and it may be necessary to make up an auxiliary focusing device, holding a screen or knife edge at precisely the distance of the camera's film plane from the end of the telescope's focusing mount.

Because of the problems in using ordinary cameras, many technically able enthusiasts make their own versions. It is only really necessary to provide a light-tight box – which can be loaded in the darkroom – and a "dark-slide" that is removed, the vibrations being allowed to die away before the exposure begins. The exposure is controlled by a "shutter" consisting of black card (or that black hat!) held in front of the telescope's aperture. As most astronomical exposures are for many seconds, minutes (or even hours), this technique is quite adequate.

Guiding

The most common method of guiding is by the use of an auxiliary telescope. This is frequently a long-focus refractor mounted on the main telescope. It is helpful if you can adjust it to pick up a bright star outside the main telescope's field if necessary. A guiding eyepiece is needed and this has a set of crosshairs – frequently glass filaments or even spider's web – or, occasionally, a reticle engraved on glass. Because high magnifications are required to ensure that tracking errors are most apparent, the sky background will be dark, and crosshairs may not be visible unless some form of faint illumination is provided.

On-axis guiding can be arranged by using a beamsplitter, which may consist of a diagonal coated with a semireflecting film, or a special prism. These divert a small amount of light away from the main beam to the guiding eyepiece.

Films and exposures

The selection of the film to be used for a particular object is largely a matter for experiment, as much depends upon the equipment and the photographer. There also has to be a compromise between the acceptable grain size, the length of exposure that can be given, and the degree of enlargement. Very fast films may allow short exposures and take advantage of short periods of good seeing, but their grain may not allow much enlargement in the darkroom, if the final picture is to remain acceptable.

When plenty of light is available, such as with the Moon, a slow-speed black-and-white film will give fine-grained negatives which can be greatly enlarged. With fainter objects, such as planets, faster films can be chosen, but many experienced photographers are prepared to retain slower-speed films for the sake of the reduced grain-size, at the expense of longer exposures, and more arduous guiding. Color films, because of their structure, are less suitable for big enlargements, and may also show color shifts in long exposures. With some films the sky background may become green, because they record the faint airglow emission. Others, less sensitive in that particular range, retain a black background. (However, many color casts can be removed either by exposing through a suitable filter – usually for a longer period of time – or by subsequent work in the darkroom.)

The eye is relatively insensitive to colors at low light levels, so photographs usually show stronger hues. Black-and-white films are insensitive to red light, and very sensitive to blue, so that many stars have photographic magnitudes that differ from the visual ones. A light yellow filter (such as a Wratten 8) will produce an approximately visual response, but requires longer exposures to reach the same limiting magnitude.

The shift in response with color films is actually an example of reciprocity failure, often mentioned as a problem with astronomical photography. Film speeds are normally calculated on the basis of exposures of a few seconds at the very most. An astronomical exposure, perhaps hundreds or thousands of times as long, will not produce a corresponding increase in the darkening of the emulsion. However, all this really means is that exposures must be established on a trial-and-error basis, rather than by following the usual relationship. With all long-exposure photography it is essential to record the exact times of the beginning and end of each exposure, as well as full details of the equipment used.

Color transparency films are particularly suitable for wide-field photographs of constellations and the Milky Way (Fig. 5.10). (Ektachrome, in particular, gives a dark background with no green cast.) Aurorae and noctilucent clouds are also very realistically rendered. The very fast films now available, although rather grainy, do allow very short exposures, and may be "push-processed" to give even higher speeds – either commercially or by the photographer. However, this again normally involves an increase in grain-size.

Color negative (print) films are not so popular, even though some very fast films are now available, but black-and-white material is extensively used for all forms of astronomical photography. Chromogenic films, with their extended range of permissible exposures, have proved to be very satisfactory for some subjects, such as star fields containing objects that differ greatly in brightness.

It is an advantage if you can process astronomical films at home, because this allows individual processing techniques to be developed. Astronomical subjects are not handled particularly well by most com-

mercial organizations, for obvious reasons, so you will have greater control if you are able to do your own processing. If you do have films commercially developed, make sure that you request them to be returned uncut, because astronomical images may confuse both automatic slide-mounting machines and human handlers. It is not unknown for automatic equipment to cut through the center of every single image in a series of photographs of the Moon, for example. In any case it is often necessary to see the edge of the frame to make certain positional measurements.

◀ Fig. 5.10 The central region of the Milky Way in Sagittarius, photographed by Chris Cook.

NEAR-EARTH EVENTS

Zodiacal light and the gegenschein

If the atmosphere is clear before dawn or after sunset, and when there is no interference from moonlight, you may be able to see the zodiacal light – a pale, tapering glow extending up into the sky (Fig. 6.1). When conditions are good, it is visible in the west after sunset, and in the east before sunrise, when its light has frequently been mistaken for the true dawn. Because it actually forms an elliptical area centered on the Sun, and with its longest dimension almost exactly along the ecliptic, it is best observed when the ecliptic is highest in the sky (Fig. 6.2). For northern observers this occurs in the west in spring, and in the east in autumn. In the southern hemisphere the seasons are correspondingly reversed. Observers in the equatorial region are particularly well placed and it can usually be seen at all seasons.

The zodiacal light is caused by scattering of sunlight by tiny interplanetary particles, most of which (like meteors) probably originate from comets, although some may be derived from minor planets. The brightest region of the zodiacal light derives from those particles that are between the orbit of the Earth and the Sun. Other particles do exist outside the Earth's orbit, but these only weakly scatter light back toward the Sun and the Earth. (They are thought to be so dark that they are unlikely to reflect much light at all.) However, at a point on the ecliptic exactly opposite to that of the Sun, you may be lucky enough to see a weak, elliptical glow known as the gegenschein. Extremely faint bridges of light also exist joining the main areas of zodiacal light and the gegenschein, but these are not readily detectable without specialized equipment, although they have been glimpsed by some keen-eyed observers under exceptionally clear, dark skies.

The main region of the zodiacal light has been compared at its brightest to that of the central regions of the Milky Way. Try photographing it with a driven or undriven camera, giving an exposure (on

► Fig. 6.1 The cone of the zodiacal light, photographed by Chris Cook. The camera has been driven to follow the sky, causing the blurred foreground.

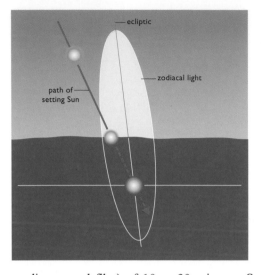

◀ *Fig. 6.2 The thin, tapering cone of the zodiacal light is best seen when the ecliptic is high above the horizon.*

medium-speed film) of 10 to 30 minutes. Such a photograph will probably show a far greater extent than was visible to the naked eye, usually being noticeably wider. However, for best results you will have to use a wide-angle lens, certainly less than 24 mm (with 35 mm cameras), if at all possible. Take care that the sky background or the morning twilight does not build up and reduce the contrast. The gegenschein has been photographed, with some difficulty, but requires exposures of at least 30 minutes. You will certainly need a fast, wide-angle lens to capture this very faint, low-contrast feature in a satisfactory manner.

It is possible to confuse the zodiacal light with certain atmospheric effects – apart from artificial lights – most especially the occasional bright glows caused when volcanic activity has injected material into the upper atmosphere. In general though, these, like the normal twilight arch, are part of a circular area centered on the Sun, as distinct from the tapering zodiacal light.

Aurorae

The Aurora Borealis (the name means "Northern Dawn") and the Aurora Australis, its southern counterpart, occur most frequently in two irregular zones, called the auroral ovals, surrounding the Earth's magnetic poles, and lying roughly between latitudes 60° and 70°. Aurorae have, however, been seen at Singapore close to the magnetic equator, so wherever you are on Earth you may hope to see a display, even if only at rare intervals. Undoubtedly many events go unrecorded because observers have not realized what they were seeing.

Aurorae are formed when electrically charged particles cascade down from the Earth's magnetosphere – its magnetic sphere of influence – into the upper atmosphere. There, the highly energetic particles excite various atoms and cause them to emit visible light. The greatest number of aurorae occur at heights of 100–115 km (around 60–70 miles), but they have been observed quite frequently as low as 70 km (43 miles), and as high as 300 km (185 miles). On very rare occasions they have been known to extend to 1000 km (620 miles). These altitudes may be compared with the approximately 80 km (50 miles) of noctilucent clouds and the 150–50 km (95–30 miles) of most meteors.

The amount of auroral activity is fairly closely linked with the sunspot cycle (page 136), but generally peaks about one or two years before sunspot maximum, when solar activity is rising rapidly. A secondary peak often occurs a year or so after sunspot maximum. Individual energetic solar flares frequently produce strong auroral displays, and a recurrence 26–28 days later is often noted, after one solar rotation, because of the persistence of particular active regions on the surface.

Aurorae occur in different forms, as described in Table 6.1. The various forms may be further subdivided on the basis of their structure and activity. Frequently observers only see the top of displays appearing over the horizon toward the pole, and this can make identification difficult. Auroral patches may be mistaken for isolated clouds, and the veils or the tops of arcs are often thought to be areas of fog. However, in general aurorae do not obscure the stars to quite such an extent as clouds or fog.

TABLE 6.1: AURORAL FORMS	
Form	Description
arc (A)	An arch-like structure, generally highest in the direction of the pole, with a fairly sharp lower edge, but a less distinct upper border.
band (B)	A ribbon-like structure, often folded like curtains, frequently showing rapid movement.
corona (C)	A set of broad rays that appear to radiate from a point high overhead.
glow (G)	A weak glow on the horizon toward the pole.
patch (P)	A faint, luminous area appearing like a cumulus cloud, often seen late in a display, frequently pulsates in brightness.
ray (R)	A narrow, vertical streak of light, frequently seen with both arcs and bands, but sometimes as isolated bundles of rays.
veil (V)	A faint, even glow of light across a large area of the sky.

Arcs and bands may be:

• homogeneous (H): without any distinct structure, but usually showing an even graduation from a bright, distinct, lower edge to a weaker, more diffuse upper border.

• rayed (R): showing distinct, vertical striations; individual rays (or bundles of rays) often extend beyond the main area of luminosity.

Auroral displays vary greatly, and can show a wide range of activity, from faint glows or patches that fail to develop further, to brilliant displays that cover much of the sky. In general, an auroral display begins quietly, with faint glows or patches in the sky toward the magnetic pole. These gradually become more pronounced, and may merge into a smooth arc across the sky (Fig. 6.4). Individual rays or bundles of rays may then appear, and the arc may develop folds, giving the appearance of auroral curtains. The folds may billow, and change shape, brightness, and color. Major displays spread across the sky, and when high overhead take on the form of a corona (Fig. 6.3). In extremely strong events the auroral ovals expand toward the equator, and observers used to seeing aurorae in the north (for example) may find that they are observing a display in the southern part of the sky. There are patterns to auroral activity, which cannot be described in detail here, but activity is normally greater after midnight.

Aurorae can be described using the abbreviations given in Table 6.1. For example, a homogeneous arc is reported as HA, and a rayed band as RB. Experienced observers also add codes to indicate colors and rapidity of motion, and include measurements of the altitude of parts of the display above the horizon. Try to obtain measurements of

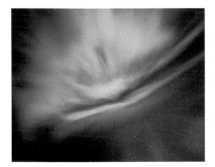

the extent of aurorae and how these change with time. An approximate estimate of the altitude of the base of an arc or band can be made by using the simple methods described earlier. It is most important to obtain the altitude and azimuth of the bottom of arcs and bands because with one observation from another site the position and height of the display can be ascertained.

▲ Fig. 6.3 An auroral corona, photographed overhead by Dominic Cantin. The two prominent streaks would appear as bands when seen from a distance.

▶ Fig. 6.4 A homogeneous arc, with some suggestion of rayed structure.

The coloration of aurorae can vary greatly, and it also depends to a considerable degree upon the observer's eyesight. Pale green and red are most often reported, but other observers of the same display may find it essentially colorless. There is also a variation in color with height, especially in long rayed structures. Color photography with fast films is likely to be of great value in providing information about the distribution in various parts of the displays.

Photography

Photography of aurorae is very rewarding. Undriven cameras are best for this work, and indeed for serious parallactic work – that is, for the determination of positions and heights – the camera should be provided with a rigid mounting allowing it to be pointed at a fixed altitude and azimuth on each occasion. This procedure greatly simplifies the process of calculation. If possible, the direction should be agreed with other observers so that the same region of the sky is covered. For a similar reason, although intermediate exposures may be made, try to take photographs beginning at exactly 0, 15, 30, and 45 minutes past each hour (UT). This allows direct comparisons to be made with those of other observers following the same pattern.

Standard and wide-angle lenses are very suitable as these usually have wide apertures and allow short exposures. With apertures of f/1.8 (or similar), exposures of 15–30 seconds on 400 ISO film (color or black-and-white) may be recommended as a starting point. If the display is very active with considerable motion of the features, you may need shorter exposures to obtain satisfactory, sharp images. Try to ensure that part of the horizon is included in any pictures as this helps to determine the exact altitude of the auroral features from your particular observing site. As with any other astronomical photographs, always record full details of equipment and all times and durations of exposures.

Noctilucent clouds

Noctilucent clouds are high-altitude, atmospheric phenomena occurring at heights of about 80 km (50 miles). They are normally observable only between latitudes 45° and 60° approximately, although it has recently been shown that they do occur at lower latitudes. At high latitudes, they appear most frequently during the weeks around the summer solstice, when twilight persists throughout the "night," and the Sun is below the horizon for the observer but can still illuminate the clouds (Fig. 6.5). This explains why they are not regularly seen closer to the equator, where suitable conditions of illumination are extremely short-lived.

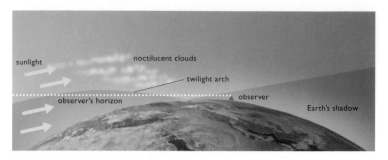

▲ Fig. 6.5 It is only during the summer months that noctilucent clouds can be illuminated by the Sun whilst the observer is in shadow.

The clouds take the form of very delicate veils, wisps, and wave-like patterns, with a silvery or bluish light, and are sometimes slightly golden toward the horizon (Fig. 6.6). The display may shift its position in the sky, with the ripples and other structures moving in a different, or even opposite, direction. Although they might seem to bear some similarities to ordinary cirrus clouds, they are about 10 times as high, and are betrayed by their appearance around local midnight, and their direction (which is generally toward the poles). Away from high latitudes, they appear fleetingly at morning and evening, when the lighting conditions are suitable. Like aurorae they are usually so thin that they do not obscure the brighter stars.

The exact nature of these clouds remains obscure. They appear to consist of tiny particles coated with ice that reflect the sunlight. It is uncertain whether the particles are meteoric dust, ions or even volcanic material injected to particularly high altitudes by violent eruptions (although it appears to be difficult for the latter to take place). The clouds occur in a very thin layer, and their different forms arise from undulations in the layer, which mean that the observer's line of sight encounters greater densities of particles in some directions than in others. The clouds' movements are related to upper-atmosphere winds, but the positions at which they occur, and some of their features, may possibly be affected by airflow over mountains far below.

Noctilucent clouds are interesting in that they appear during the period when auroral observation is most difficult. However, they may be studied with almost identical methods, both visually, when they are classified into a number of different types (Table 6.2), and also photographically. Make observations at suitable intervals – say every 15 minutes – recording the changes in appearance and general motion of the clouds that take place. Measurements of the angles to various parts of the display are easy to make.

TABLE 6.2. NOCTILUCENT CLOUD TYPES	
Type I: Veils	Tenuous sheets that resemble cirrus or cirrostratus clouds, occasionally with a slightly fibrous structure. This type often occurs before other types, or as a background form.
Type II: Bands	Elongated streaks that often lie parallel to one another, but sometimes cross at a shallow angle. They may alter in brightness, generally over periods of 20–60 minutes.
Type III: Billows	Short streaks that are closely spaced and approximately parallel to one another. They occur on their own or with the longer bands. They are often variable, changing shape and brightness over periods of a few minutes.
Type IV: Whirls	Partial or complete rings or loops of cloud, surrounding darker centers. Complete (i.e. closed) rings are only rarely seen.

Photography

Photographic techniques for noctilucent clouds are similar to those used for auroral observations – ideally photographs are taken with fixed cameras covering the same region of sky as other observers, and making a series of exposures at fixed intervals. Because these clouds are brighter than most aurorae, you can either use slower films with the advantage of fine grain or else choose smaller apertures. Exposures may have to be reduced, not because of rapid motion, but because otherwise the bright sky background might fog the film and reduce the contrast. Color photography is particularly effective, some films such as Kodachrome giving excellent rendering very similar to the visual appearance. With f/2 lenses and 100 ISO film, you might like to begin by making exposures of 5, 3, and 1 seconds. As with aurorae, aim to start a series of exposures at the exact quarter-hours (that is, at 0, 15, 30, 45 minutes past the hour).

▲ *Fig. 6.6 A fairly typical display of noctilucent clouds, with bands and billows, photographed by Neil Bone from West Sussex.*

Meteors

When particles or small bodies orbiting the Sun plunge at high speeds into the Earth's atmosphere, ionizing the atoms, they give rise to the streaks of light known as meteors ("shooting stars") (Fig. 6.7). Occasionally these may be very brilliant, if the particles are particularly large. When they are brighter than about magnitude -5, somewhat more than the maximum that Venus can reach, they are called fireballs. The meteoroids (as the particles are termed when they are out in interplanetary space) may be completely vaporized or may disintegrate during their passage through the atmosphere. If, however, they were sufficiently large when they entered the atmosphere, fragments may survive to fall on to the surface of the Earth, when they are known as meteorites. In general, any meteorite will have produced a brilliant fireball during its descent. The recovery of these objects is obviously very important. As yet they are the only celestial bodies that we can examine, apart from the samples returned from the Moon. Even if you are not particularly interested in the fainter meteors, you should at least know what information to record if you observe a fireball.

Observing meteors

You can certainly see meteors on any clear night, and if conditions are good and you are observing with the naked eye, expect there to be 5–10 in an hour. These are sporadic meteors, individual particles orbiting alone, which can appear at random, quite unexpectedly, anywhere in the sky. However, at certain times of the year, when an individual meteor shower is active, meteors are far more frequent. Such showers

are caused by groups of particles, traveling together along an orbit which, at some point, intersects that of the Earth. It is only while the Earth is close to that particular portion of its own orbit that the members of the shower are seen. Many showers can be linked with known or extinct comets, such as the Leonids with Comet Tempel-Tuttle, and the Orionids with Comet Halley. It is believed that most of the sporadic meteors have the same general

◄ Fig. 6.7 Bright Leonid meteors are seen against a backdrop of Orion.

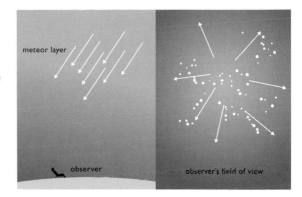

► Fig. 6.8 Meteors enter the atmosphere on parallel paths but perspective causes them to appear to diverge from the radiant.

source of origin. The major showers are listed in Table 6.3, but it should be noted that the rates vary considerably from year to year, and bright moonlight also greatly reduces the number of meteors that are visible. At present, the most reliable showers are the Quadrantids, Perseids, Leonids, and Geminids, and keen observers make every attempt to watch these events.

Although shower meteoroids follow parallel paths in space and in the upper atmosphere, perspective makes their tracks appear to diverge from a single area of sky, known as the radiant (Figs. 6.8 and 6.9). Showers are generally named after the constellations where their radiants are found, and some of the most important are listed in the table. However a few meteor showers have been named after associated comets, and the most important of these that you might

▲ Fig. 6.9 A photograph taken by Chris Cook of the Leonid meteor shower radiant, showing several individual meteors.

TABLE 6.3: METEOR SHOWERS						
Shower	Maximum	Normal limits	Rate at maximum	Radiant RA	Dec.	Remarks
Quadrantids	Jan 04	Jan 01–06	100?	15^h 28^m	+50°	Blue meteors with trains
Lyrids	Apr 22	Apr 19–25	10–15	18^h 08^m	+32°	Bright meteors
η-Aquarids	May 05	Apr 24–May 20	40	22^h 20^m	−01°	Broad maximum and multiple radiants
α-Scorpids	Apr 28	Apr 20–May 19	10	16^h 32^m	−24°	Multiple radiants –
	May 12			16^h 04^m	−24°	long activity April to July
δ-Aquarids	July 28	July 15–Aug 20	20	22^h 36^m	−17°	Double radiant – southern component
	Aug 06			22^h 04^m	+02°	is richer
Perseids	Aug 12	July 23–Aug 20	75	03^h 04^m	+58°	Rich shower, bright meteors with trains
Orionids	Oct 21	Oct 16–26	25	06^h 24^m	+15°	Fast meteors, many with trains; associated with Comet Halley
Taurids	Nov 03	Oct 20–Nov 30	10	03^h 44^m	+14°	Slow meteors, many fireballs
Leonids	Nov 17	Nov 15–20	100?	10^h 08^m	+22°	Trains, rates declining after major storms in 1999–2002
Puppids-Velids	Dec 08	Late Nov–Jan	15	09^h 00^m	−48°	Two of several radiants
	Dec 25			09^h 20^m	−45°	
Geminids	Dec 13	Dec 07–15	100	07^h 28^m	+32°	Many fireballs
Ursids	Dec 22	Dec 17–25	10	14^h 28^m	+78°	Increased activity in certain years

come across are probably the Bielids (or Andromedids) and the Giacobinids (or Draconids).

Meteor numbers are always calculated as hourly rates. Those given in Table 6.3 are only approximate figures and might apply if you were an experienced observer, watching the region of the zenith under very good conditions. Meteors closer to the horizon are dimmed by atmospheric extinction, and general seeing also plays a part. Moonlight causes grave interference for five or six days before and after Full Moon, so in some years, individual showers may be well-nigh impossible to observe. In addition, the numbers vary from year to year, either gradually, or dramatically, depending upon whether the particles are spread more or less evenly around the orbit, or are concentrated into dense clumps.

The Leonid shower is the most striking example of this annual variation. There were high rates in 1799, 1833, and 1866 (and probably earlier in history), but disappointing numbers in 1899 and 1932 when the gravitational effects of Jupiter and Saturn moved the orbit away from the Earth. In 1966 there was an astounding display, when the hourly rate rose to about 140,000 for a period of around 20 minutes. There have been major Leonid storms in recent years. In 1999 the hourly rate was about 2500–3000; and in 2000 it had a broad plateau of about 200–250, with two peaks, one reaching about 750. In 2001, rates were about 2500, again with two peaks, one about 4000; and in 2002 the maximum was about 2800. Leonids rates are now expected to decline rapidly. It must be added hastily that such numbers are quite exceptional and that only three of the normal showers (the Quadrantids, Perseids, and Geminids) can be expected to exceed rates of 50 per hour.

The number of meteors observed also changes throughout the night. Before midnight the only meteors seen are those that are overtaking the Earth, and have slow apparent velocities (Fig. 6.10). After midnight the velocities of the Earth and meteors combine, and because the velocity strongly affects the brilliance, the number observed increases after midnight.

Visual observations

The best method of observing meteors visually is for a number of observers to form a group and watch together. Each observer can then cover part of the sky, and one member of the team can act as recorder for the others. However, it is still well worth trying even if you have to observe on your own. Because meteors appear at random intervals you do need to be prepared to watch continuously for 30 minutes at a time. (Watches should always last for multiples of 30 minutes, but it is usually just as well to have a break in between.) Make sure that you have enough clothing, because you soon become chilled when sitting

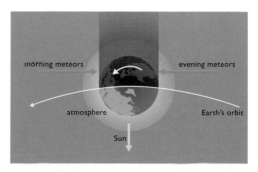

▶ Fig. 6.10 Because of Earth's rotation and orbital motion, more meteors are seen after midnight, when they have higher velocities.

(or lying) still for that length of time. Keep a note of the times at which your watches begin and end.

Where should you look? About 45° from the radiant is best, and as always, as high in the sky as possible. One observer cannot hope to cover much of the sky, so do not worry about what might be happening behind you. Obtain a set of charts that you can mount on card and cover with transparent plastic film. (You will usually only need the one showing the part of the sky that you are watching.) If you can, roughly estimate the magnitude of the faintest star that you can see at the beginning and end of each watch. This gives information about the sky conditions so that your observed rates can be corrected as necessary.

Ideally the following details would be recorded for each meteor: time, path, type, brightness, and any special features. If you are observing when a shower is very active it may not be possible to do this for every meteor. The last three pieces of information are then the most important. Let's take the details in order, omitting time, which rarely presents any problems.

Path Recording the path is not too difficult. When you see a meteor, hold a piece of string, or even better a straight stick, along the path to help to fix its position against the stars. Estimate the start and end points (and another in between if possible). You might say "a third of the way from γ to α Leonis, over ι Leo, and halfway between δ and γ Vir," for example. Draw the track on the chart. A word of warning here. Meteor paths can be shown as straight lines only on charts with a special form of projection, known as gnomonic projection. Such charts are not easy to obtain and are difficult for inexperienced observers to use because constellation patterns are distorted and they do not look much like the sky. Meteor tracks on other projections are curved lines, but if you can plot the two end points accurately, any information about the track and orbit can be calculated if necessary. If you have covered the chart with plastic and use a non-permanent marker, you can wipe off the tracks when you have transferred all the information into your observing notebook. During a major shower just note the constellation in which the meteor was seen.

Type of meteor Next decide if the meteor belonged to a shower or was sporadic. This can be done by either "sliding" the stick back along the trail, or mentally projecting the line, to see if it comes from the radiant of any shower active that night. If the line passes within 4° of the position of a radiant, you can safely assume that the meteor belonged to that shower. Before you start observing, draw the position of the radiant on your chart. (A radiant slowly moves as the Earth passes through the meteor stream. Information about the daily motion can be found in one of the astronomical yearbooks.)

Brightness Estimates of brightness give a lot of information about the size and velocity of the particles. Unlike variable-star methods (page 177) an accuracy of about half a magnitude is all you can hope to obtain. This is not too difficult because you can usually compare a meteor's magnitude to that of a star in the area, or say it was halfway between one and another. Do not try to remember lots of magnitudes, just note down the names of the stars (or mark them on the chart), and look up the values afterward. Try to choose stars close to the meteor's track, so that the extinction is about the same. Very bright meteors can be a problem as there may be no suitable comparisons, and you will just have to rely upon your memory of stars such as Sirius (magnitude −1.4), or how bright Jupiter and Venus may become (magnitudes −2.4 and −4.3, respectively).

Special features Some meteors give rise to persistent luminous trains, lasting for several seconds or even longer. Make a note of their duration, changes of shape and position. They are not common, so any observations are valuable. (The changes in shape and position, for example, provide information about winds in the upper atmosphere.) Other special information may include notable color and terminal bursts. These features are usually only seen with the brighter meteors.

Photographic observations

Meteor photography is not difficult, but does require patience, as only the brighter meteors are recorded, and many exposures may have to be made before one is captured. Modern fast films and wide-aperture lenses help greatly in this respect. Cameras should be undriven as background star trails will not matter, provided they can be identified. Like visual work, the best area to survey is about 45° from the radiant. Some dedicated observers arrange several

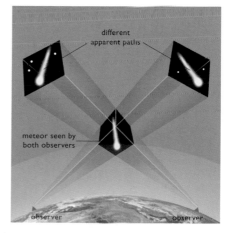

▶ *Fig. 6.11 Observation (or photography) of meteors from two different sites allows their heights and paths to be determined.*

cameras so that they cover the whole of the sky, others invest in fish-eye lenses. It is quite a good idea to observe the same area by both visual and photographic methods, as then the times of any bright meteors recorded by the camera will be known accurately. Photographs allow information about magnitudes and positions to be obtained with reasonable ease, and have the advantage of providing a permanent record.

If two cameras many miles apart are arranged so that their fields cover the same volume of the upper atmosphere, meteor triangulation becomes possible and the exact track of the meteor and its heights can be determined (Fig. 6.11). (Visual observers sometimes cooperate to get similar results.) If possible, the meteors are photographed through a rotating-sector shutter (looking like the blades of a fan), operating in front of the camera lens. This breaks the trail into segments (perhaps 10 a second), from which the actual velocity of the meteor may be derived. In this way it may even be possible to find the precise orbit that it had in space. Changes in the spacing of the segments of the trail show how the body was braked by the upper atmosphere, and can lead to a determination of its density.

Telescopic work

Deliberate meteor watches with binoculars or telescopes are only really suitable for very dedicated observers. The fields of view are restricted, but fainter meteors become visible, giving information about the smaller sizes of meteoroids. However, if you are observing something else – variable stars, galaxies, etc. – you may well see the occasional meteor. Try to record the necessary information about direction, magnitude, color, and speed, or better still, make a quick sketch of the field and track (Fig. 6.12). Any observations are valuable because many are needed before proper analysis becomes possible.

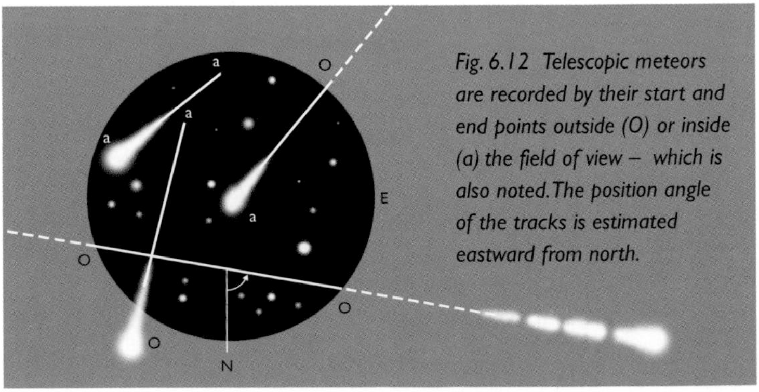

Fig. 6.12 Telescopic meteors are recorded by their start and end points outside (O) or inside (a) the field of view – which is also noted. The position angle of the tracks is estimated eastward from north.

Fireballs

Any meteor brighter than magnitude −5 is known as a fireball. Some fireballs may be exceptionally brilliant, ranging beyond magnitude −15. (The Full Moon is magnitude −13.) If they are seen at night all the usual techniques apply, but duplicate photographs are particularly valuable because they help to determine whether any meteorite may have fallen to Earth, and where this is likely to be.

Very occasionally fireballs are so brilliant that they can be seen in daylight. Any observations are then of outstanding importance. If you should be lucky enough to see one, make a note of the time, make a guess at its brightness, and establish its path. You will not have stars to act as reference points, but you can estimate the altitude and azimuth of the beginning and end of the path, or refer to landmarks on the ground. If you can, establish your exact position and make a note of that. Then wait. The largest fireballs are sometimes accompanied by audible sounds – when they are known as bolides – and these may take several minutes to reach you. There may be a sonic boom because the body is traveling faster than the speed of sound and, in addition, there may be explosive fragmentation of the main body. Sometimes the noise produced in this way can be quite prolonged. If you note the times at which you hear any noises, the distance to the track can be established. Report any details to your national fireball organization immediately, who may well send an investigator to check some of the details with you, especially if a meteorite fall might be involved. Fireballs may sometimes be confused with satellite re-entries, but there are ways in which they may be distinguished from one another (page 115).

Artificial satellites

There are so many objects in orbit around the Earth that observing sessions in the early evening or before dawn rarely pass without some satellites being visible as they pass across the sky. ("Satellites" may, of course, be taken to include spent upper-stage rockets, shrouds, and miscellaneous bits and pieces as well as the active satellites themselves.) Although some objects can be followed right across the sky, many disappear or reappear as their paths take them into, or out of, Earth's shadow.

Many of these objects are rotating or tumbling in their orbits, and show distinct flashes or changes in brightness as sunlight is reflected from large flat surfaces (such as solar panels) and other parts of the structure. Because of their design, the numerous Iridium communications satellites are particularly noticeable (Fig. 6.13). Their brightness slowly increases and, at its peak, the flare may be similar in magnitude to Venus at its brightest, after which the brilliance declines at a similar

◀ Fig. 6.13 A typical, brilliant flash from one of the many Iridium satellites, photographed by Dave Sewell.

rate. The Mir space station was prominent for many years, and the International Space Station has become brighter as various modules were added over the years.

It is only when the observer is in shadow and the satellite in sunlight that any object can be seen. As a result, the periods of visibility vary with the observer's latitude and the time of year, and also depend on the satellite's orbital height and inclination. In summer, a high-orbit satellite may be visible at any time during the night for an observer at high latitudes. At other times, the same satellite may be visible for only a very short period low on the horizon. This means that prediction of the times when individual satellites are visible from any place is fairly complicated. It can be carried out by those who like such calculations, but most observers rely upon precise predictions issued by national coordinating bodies. (Approximate times of appearance for a few bright objects are given in some newspapers, but more extensive, and extremely accurate details, including dates and times of predicted re-entries, are available over the Internet.) Once an expected track has been plotted on a star chart, observations can be made with binoculars or larger telescopes, the former being most suitable for the majority of amateurs.

Serious observing involves defining the position of a satellite at a particular time, determined by a stopwatch or other means. The most accurate method is to note when the satellite passes between two stars, but this is not always possible, and other determinations sometimes have to be used (Fig. 6.14). At least two positions are needed on each pass for the orbit to be determined. Comparison of the predicted and observed paths then allows deductions to be made about the density of the upper atmosphere (particularly where the satellite is closest to the Earth, at perigee), which fluctuates with solar activity, and also about the exact size and shape of the Earth. Observations of the magnitude and flash rate give information about the condition of the satellite itself.

Satellites move so slowly that they are rarely taken for other objects, but re-entries may be confused with meteors or even bright fireballs. You can usually tell the difference by noting their apparent velocities and directions. Satellites orbit and re-enter slowly, at speeds of 4–8 km (2.5–5 miles) per second, whereas the theoretical absolute minimum for a meteoroid is about 11 km (7 miles) per second, and most have speeds higher than this – up to about 70 km (43 miles) per second. Observing meteors will give you an idea of what apparent speed may be expected.

The direction in which an object is traveling may also provide a useful clue. At present no manned flights occur at high orbital inclinations, and re-entry, or the re-entry of carrier rockets or fuel tanks, carries them from west to east. Polar-orbiting satellites have much higher inclinations and cross the sky in a general north–south or south–north direction. In addition, satellites frequently break into fragments when they re-enter and produce multiple trails. (Some may produce extensive showers of debris.) This is fairly uncommon among natural meteoroids, although some of the larger fireballs may fragment and produce multiple trails.

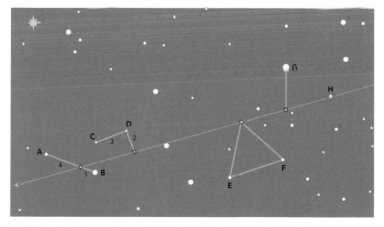

▲ Fig. 6.14 A satellite's position is given by the ratio when it crosses a line between two stars (A and B) or is at right angles to a pair (C and D). Other positions are when it forms an equilateral triangle with two stars (E and F), is vertically above or below a star (G) or passes close to another (H).

THE MOON

The Moon is usually the first object that people examine when they start to become interested in astronomy. It seems the obvious choice. It is large (about 30′ in diameter), bright, and easily found. Its brightness, too, means that it is quite possible to observe the Moon in daylight. In fact, this is an excellent idea, because it reduces the amount of glare from the surface, and is quite satisfactory provided the seeing conditions are reasonably steady. If a telescope is used at night, it may be necessary to use a neutral-density filter or else reduce the telescope's aperture to provide more comfortable observing conditions.

Phases of the Moon

The motion of the Moon is extremely complex and the calculation of the exact times of Moonrise and Moonset is very complicated, especially because the observer's position on Earth must be incorporated. You will probably find that the details given in many newspapers are quite adequate, but if you need very accurate information you will have to use one of the astronomical yearbooks or an appropriate computer program. On average, the Moon rises and sets each day about 50 minutes later than the day before, but the amount varies very greatly.

A complete cycle of the familiar lunar phases from one New Moon to the next takes approximately 29.5 days, or one lunation (Fig. 7.1). The Moon's age is reckoned from the time of New Moon, when it is, of course, closest to the Sun in the sky, or may even pass in front of its disk in a solar eclipse. Many people enjoy the challenge of trying to spot the hair-thin crescent when it is merely a few hours old, and the technique suggested for locating Mercury may prove useful (page 147).

Just as the Sun's elevation changes throughout the seasons, so too does that of the Moon. As a result the best time for studying a particular phase falls at a specific time of year. The Full Moon is best examined in midwinter, when the Sun is lowest and the Moon highest. The crescent phases around New Moon, on the other hand, may well be better at midsummer. For studying First Quarter, northern-hemisphere observers should choose the spring (autumn for those in the south). Similarly Last Quarter is best in autumn for northern astronomers and in spring for southern observers. The Moon's motion 5° on either side of the ecliptic has an additional effect upon visibility at these particular times. Observers in the tropics are lucky in that they are able to observe at any time of the year, but then the periods of visibility are not so long.

Lunar features

Some lunar features may be seen even by the naked eye and, as mentioned earlier, you will find that it really is worthwhile trying to make a drawing of these without any optical aid, particularly as training for planetary observation. The dark areas may be easily outlined, but sharp-eyed observers can also make out some of the details in the brighter portions, especially when the changing shadows throw some parts into relief.

Binoculars begin to show that the dark and light areas are very different. The bright regions prove to be rugged, cratered highlands (occasionally known as *terrae*), quite distinct from the lower, smoother, dark plains. The Latin term *mare*, or "sea" (pl. *maria*), is still retained for most of these low-lying areas as a reminder of the days when they were

▼ *Fig. 7.1 The phases of the Moon, beginning shortly after New Moon.*   North is at the top, matching the naked-eye view from the northern hemisphere.

TABLE 7.1: LUNAR FEATURES		
Term	Feature	Example
Dorsum (Dorsa)	Ridge(s)	Dorsa Smirnov
Mare (Maria)	"Sea" – flat lava plain(s)	Mare Humorum
Mons	Mountain	Mons Pico
Montes	Mountain range	Montes Altai
Palus	"Swamp" – irregular dark plain	Palus Putredinis
Rima(e)	Rille(s) or cleft(s)	Rima Ariadaeus
Rupes	Fault	Rupes Recta
Sinus	"Bay" – mare area	Sinus Iridum
Vallis	Valley	Vallis Alpes

wrongly thought to be bodies of water. The largest of the craters appear fairly distinct through binoculars, but even the smallest telescopes reveal a far greater wealth of detail. Craters of all shapes and sizes are visible with a 75 mm telescope as well as many other features such as the valley-like rilles, ridges, isolated peaks, and extensive mountain ranges. Like many other astronomical objects, lunar features have internationally known, official, Latin names, and these are used here. The general terms for the various types of feature are listed in Table 7.1.

The continual variation in the elevation of the Sun means that the appearance of the features and their shadows are also changing all the time. Many of the details are seen most clearly when they are close to the terminator (the line dividing the illuminated and non-illuminated portions of the surface). It is then, at sunrise and sunset, that the low angle of the lighting causes everything to appear in exaggerated relief.

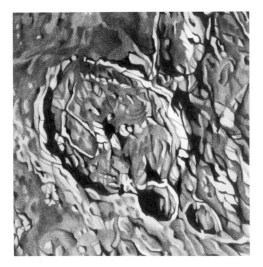

◀ Fig. 7.2 An observational drawing of the crater Posidonius by Peter Grego.

▶ *Fig. 7.3 The Full Moon, with north at top, as seen by the naked eye and through binoculars by observers in the northern hemisphere. Photograph by Mike Goodall.*

In fact, the Moon's rugged appearance is mostly an illusion caused by the dramatic, high-contrast lighting. Most of the slopes on the Moon are very gentle and are much less steep than those found on the Earth.

Under high illumination, some features, even very large craters, become indistinct or virtually disappear, or may only be seen by virtue of the different amount of light that they reflect when compared with the surrounding surface. A few features may become more conspicuous because of this effect. The most notable examples are the ray systems extending from some of the craters, such as Tycho, Copernicus, and Kepler. These are hardly visible at early and late phases, but are most conspicuous under a high Sun, around the time of Full Moon (Fig. 7.3)

The Earth itself, being covered with clouds and areas of ice, reflects a considerable amount of sunlight back into space. It is said to have a high albedo. When the Moon is a thin crescent, a few days before or after New Moon, this reflected "Earthshine" can be seen illuminating the portion of the surface otherwise in shadow. Some of the lunar features (especially craters such as Aristarchus, Kepler, and Copernicus) themselves have high albedos, and may be distinctly seen by this dim reflected light.

▼ *Fig. 7.4 Prominent Earthshine, just after New Moon, photographed by Stephen Pitt.*

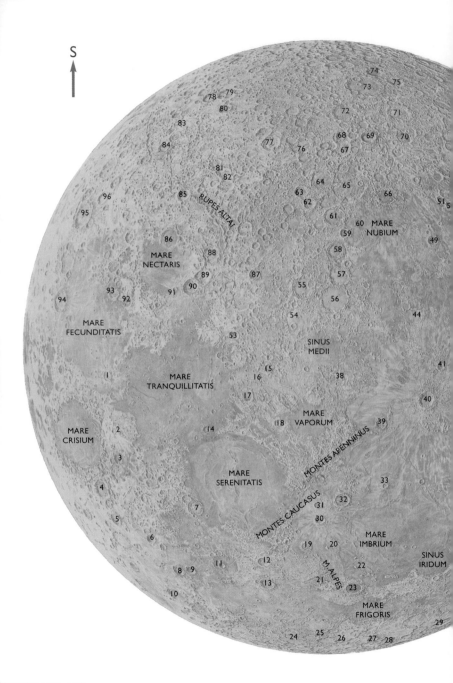

S

MARE NUBIUM

MARE NECTARIS

RUPES ALTAI

MARE FECUNDITATIS

SINUS MEDII

MARE TRANQUILLITATIS

MARE VAPORUM

MARE CRISIUM

MONTES APENNINUS

MARE SERENITATIS

MONTES CAUCASUS

MARE IMBRIUM

SINUS IRIDUM

M. ALPES

MARE FRIGORIS

Key to Moon map (south at top)

The two approximate ages of the Moon when a feature is best seen are indicated by the figures following the name. These dates may vary by a day, owing to libration and other factors.

1	Taruntius	4:18		49	Bullialdus	9:23
2	Proclus	14:18		50	Campanus	10:24
3	Macrobius	4:18		51	Mercator	10:24
4	Cleomedes	3:17		52	Schickard	12:26
5	Geminus	3:17		53	Delambra	6:20
6	Franklin	4:18		54	Hipparchus	7:21
7	Posidonius	5:19		55	Albategnius	7:21
8	Atlas	4:18		56	Ptolemaeus	8:22
9	Hercules	5:19		57	Alphonsus	8:22
10	Endymion	3:17		58	Arzachel	8:22
11	Bürg	5:19		59	Thebit	8:22
12	Eudoxus	6:20		60	Birt	8:22
13	Aristoteles	6:20		61	Purbach	8:22
14	Plinius	6:20		62	Werner	7:21
15	Agrippa	7:21		63	Aliacensis	7:21
16	Rima Ariadaeus	6:20		64	Walther	7:21
17	Julius Caesar	6:20		65	Deslandres	8:22
18	Manilius	7:21		66	Pitatus	8:22
19	Cassini	7:21		67	Orontius	8:22
20	Mons Piton	8:22		68	Saussure	8:22
21	Vallis Alpes	8:22		69	Tycho	8:22
22	Mons Pico	8:22		70	Wilhelm	9:23
23	Plato	8:22		71	Longomontanus	9:23
24	Meton	6:20		72	Maginus	8:22
25	Barrow	7:21		73	Clavius	9:23
26	Anaxagoras	9:23		74	Blancanus	9:23
27	Philolaus	9:23		75	Scheiner	10:24
28	Anaximenes	11:25		76	Stöfler	7:21
29	Pythagoras	12:26		77	Maurolycus	6:20
30	Aristillus	7:21		78	Vlacq	5:19
31	Autolycus	7:21		79	Hommel	5:19
32	Archimedes	8:22		80	Pitiscus	5:19
33	Timocharis	8:22		81	Rabbi Levi	6:20
34	Aristarchus	11:25		82	Zagut	6:20
35	Herodotus	11:25		83	Janssen	4:18
36	Va. Schröteri	11:25		84	Metius	4:18
37	Mons Rümker	12:26		85	Piccolomini	5:19
38	Pallas	8:22		86	Fracastorius	5:19
39	Eratosthenes	8:22		87	Abulfeda	6:20
40	Copernicus	9:23		88	Catharina	6:20
41	Reinhold	9:23		89	Cyrillus	6:20
42	Landsberg	10:24		90	Theophilus	5:19
43	Kepler	10:24		91	Mädler	5:19
44	Fra Mauro	9:23		92	Gutenberg	5:19
45	Grimaldi	13–14:27–28		93	Goclenius	4:18
46	Letronne	11:25		94	Langrenus	3:17
47	Billy	12:26		95	Petavius	3:17
48	Gassendi	11:25		96	Snellius	3:17

Libration

The Moon always turns the same face to the Earth, and at first this may seem to be unvarying. Closer attention over a period of a few lunations shows that alterations in the visibility of the features do take place. These are the result of the effects known as libration, which make the Moon appear to rock slowly backward and forward before our eyes.

The inclination of the Moon's orbit takes it above and below the plane of the Earth's equator, causing libration in latitude so that we alternately see more of the northern hemisphere and then more of the southern. Because of the Moon's elliptical orbit around the Earth its speed varies quite considerably. It rotates on its axis at a constant rate, however, and because the two do not match all round the orbit, libration in longitude results. Parts of the farside appear and disappear over the eastern and western limbs. (The limb is the apparent edge of any body that shows a distinct disk.)

Various other libration effects also combine with those mentioned to expose about 59% of the surface to view over a very long (30-year) period. Some of the formations on the edge of this visible region are very rarely seen under favorable lighting conditions. Because of libration, the terminator does not sweep across the various formations with monotonous regularity every month, but may reach a particular feature as much as half a day earlier or later than average.

There is also a change in the Moon's size between the nearest and farthest points in its orbit (perigee and apogee) which, although not apparent to the eye, shows on photographs taken at those times and given the same degree of enlargement. However, unlike libration, this has little effect upon observations.

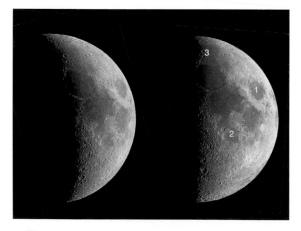

◄ Fig. 7.5 The effects of libration are demonstrated in these two images, taken on 2001 March 1 (left) and 2000 June 8 (right). Mare Crisium (1) is clearly farther away from the limb in the image at right; Mare Nectaris (2) is farther north and west; and the line of craters near the top of the disk (3) are much nearer to the north pole.

The surface of the Moon

Start to observe the lunar features by following the passage of both the sunrise and sunset
terminators across the disk, watching the
individual formations appear and disappear in
turn. A guide to the position of the terminator
and to the visibility of individual formations
on any particular day of the lunation is given
in the list that accompanies the Moon map
(pages 120–121).

The features at the limbs are difficult to
observe around the time of New Moon, so
they must be studied just before and after it is
Full. These are, of course, the regions most
affected by libration, so individual features
given in the list may not always be visible.
Because of the effects of foreshortening, it is
frequently very difficult to interpret the details
at the limbs, and positive identification of
features may prove to be impossible. Only
repeated observation will enable this to be
done with any degree of confidence.

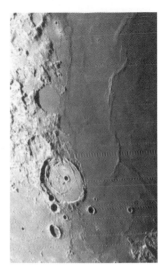

▲ Fig. 7.6 Part of
Mare Serenitatis,
showing Posidonius
and the wrinkle ridges
called Dorsa Smirnov.

Maria

We now know that the dark mare areas of the Moon have been
flooded with lava (Fig. 7.6). In some cases (the circular maria) lava
has filled basins excavated by the impact of large meteoroids. Mare
Imbrium is the most notable example, but others are Mare Serenitatis,
Mare Humorum, and Mare Crisium. Other maria are irregular and
less well defined; Mare Frigoris and Mare Vaporum are good examples.
Oceanus Procellarum is so vast, covering over 2,000,000 sq km
(800,000 sq miles), that it is really in a class by itself. Several of the
maria exhibit extensive wrinkle ridges, the most prominent being
those in Mare Serenitatis and Oceanus Procellarum, although there are
others in Mare Tranquillitatis. (The last region is probably best known
as the site of the first Moon-landing.)

Mountain ranges

Some of the Moon's most conspicuous mountain ranges form the
edges of maria. The Montes Carpatus, Apenninus and Alpes (with
the distinctive Vallis Alpes) around Mare Imbrium, and the Montes
Caucasus bordering Mare Serenitatis are very notable (Fig. 7.7). The
Rupes Altai is rather different, and appears to be an old formation that

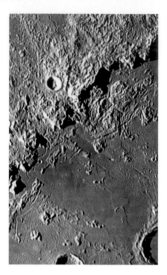

▲ Fig. 7.7 Part of
Mare Imbrium, with
Montes Apenninus,
Palus Putredinis, and
Archimedes.

has been degraded and partially overlaid by later deposits. Individual isolated peaks are also encountered such as Pico and Piton in Mare Imbrium and the ill-defined mass of Mons Rümker at the junction of Sinus Roris and Oceanus Procellarum. The closest that the Moon comes to distinct volcanic structures are the low domes, such as those near the craters Arago and Hortensius.

Craters

Impacts by meteoroids were responsible for the most conspicuous characteristic of the Moon – its craters. They cover a vast range of sizes. Some of the largest are well defined, such as Ptolemaeus, Schickard, and the enormous but rarely visible crater Bailly, 300 km (185 miles) in diameter. Others, for example Hipparchus and Fra Mauro, are very degraded. Many craters, such as Copernicus and Theophilus, have terraces on their interior walls, while central peaks are fairly common, as in Petavius, Eratosthenes, Copernicus, Aristillus, and Theophilus. There are innumerable smaller, fairly regular, bowl-shaped craters, down to the limit of visibility. In the lunar highlands craters frequently overlap and break into one another. Thebit has a well-formed small crater (Thebit A) centered on its wall, and is probably the best example. Some smaller craters are very distinctive, such as those in the crater chain on the floor of Clavius, and the dark-haloed pits in the floor of Alphonsus. Distinct internal details are seen in some craters, perhaps the most famous being the dark bands within Aristarchus.

Secondary craters and rays Secondary craters are formed when material thrown out by the main impact itself excavates smaller craters. Such patterns and blankets of ejecta are distinctly visible, especially around Copernicus and Bullialdus. Other, finer ejecta have produced the pale ray-systems spreading out over the surface, and visible at Full Moon. Tycho's system is the most extensive, and the crater itself is a major feature with a central peak and a definite dark halo (Fig. 7.8). Copernicus, Kepler, and Aristarchus also have conspicuous rays, while those of Proclus are very asymmetrical. In Mare Fecunditatis, the pair of craters Messier and Messier A present a comet-like appearance with their one prominent ray. It is thought that they were formed by a low-angle impact that threw most of the ejecta in a single direction.

Flooded craters Some craters have very dark floors, indicating flooding by lava. Endymion, Archimedes, Plato, and Grimaldi are some of the most notable examples. Some craters have quite obviously been breached by mare lava flows, and in the case of Sinus Iridum this is particularly conspicuous (Fig. 7.9). Prinz, Letronne, and Fracastorius are some other features of this type. The interior of Wargentin has been flooded up to the level of the rim, giving rise to a plateau. There are large numbers of ruined formations, but a few are truly called "ghost" craters. These are old features apparently completely covered by flows of lava, which on cooling has contracted, producing a mere suggestion of the underlying structure. Stadius in Sinus Aestuum between Eratosthenes and Copernicus is the best example, but another is south of Lambert in Mare Imbrium, and Lamont in Mare Tranquillitatis is rather more prominent.

Rilles

Other common types of formation are the various kinds of rilles and clefts (which are all given the general Latin term *rima*). Some of these appear to be directly related to the formation of individual features, such as the clefts that run parallel to the borders of Mare Humorum.

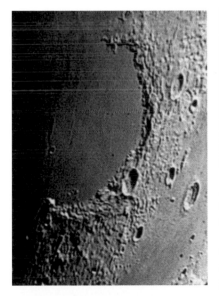

▲ *Fig. 7.8 The crater Tycho has a distinct dark halo and is the source of the largest ray system.*

◀ *Fig. 7.9 Sinus Iridum is the remnant of a crater that has been flooded by lava from Mare Imbrium.*

◀ *Fig. 7.10 The Triesnecker (center) and Hyginus (bottom left) rilles in Sinus Medii.*

Rilles, which may be less obviously structural, are prominent in the Triesnecker, Ariadaeus, and Hyginus systems, all of which are in the area bounded by Sinus Medii, Mare Vaporum, and Mare Tranquillitatis.

Sinuous rilles are rather different, and owe their winding nature to the fact that they formed from lava tubes, the roofs of which later collapsed. Vallis Schröteri (Schröter's Valley), close to Aristarchus, is possibly the best example, while the similar Prinz rilles are nearby. The great rift of Vallis Alpes (the Alpine Valley) has a narrow, sinuous rille on its floor, and Rima Hadley, the site of the Apollo 15 landing, is another good example.

Rifted floors occur in some craters, such as Gassendi, Petavius, and Posidonius, while a rift passes through both walls and floor of Goclenius, continuing the line of a cleft outside the crater. The most distinct single fault is the Rupes Recta (often known as the "Straight Wall") in Mare Nubium – very striking at sunset, but in reality with quite a gentle slope of only about 7°. Rupes Cauchy is more like a cleft, but casts a conspicuous shadow only at sunrise.

Transient phenomena

There are particular areas of the Moon that occasionally seem to show changes in brightness. These Lunar Transient Phenomena (LTPs) are of uncertain origin, but may be related to solar activity, or to the occasional escape of gas from the lunar interior. They cause short-lived color changes to the areas affected, and can be studied by examining the surface through red and blue filters. The simplest device – called a "Moon-blink" – allows these to be switched rapidly backward and forward, when any change in brightness becomes immediately apparent. Some areas, such as Fracastorius show permanent "blinks," but others, particularly Aristarchus, Gassendi, and Alphonsus, are the sites of definite activity.

Lunar eclipses

Eclipses of the Moon occur when it passes through the Earth's shadow. This can happen only at Full Moon, but because of the inclination of the orbit to the ecliptic, on most occasions the Moon does not encounter the shadowed zone. However, two or three eclipses occur every year and each one is visible to any observer on the hemisphere facing the Moon. The motion of the Moon through the shadow is always from west to east (as seen against the background stars). Some forthcoming lunar eclipses are listed in Table 7.3.

There are two regions to the Earth's shadow: an inner dark cone pointing away from the Earth (the umbra), and a wider, less dense cone with its apex at the Sun (the penumbra). Although penumbral eclipses occur if the Moon's path crosses just this region, they are of

Date		Type	Region of visibility
TABLE 7.3: FORTHCOMING LUNAR ECLIPSES			
2012	Jun 04	Partial	Pacific, E. Australia
2013	Apr 25	Partial	W. Asia, India, Africa, E. Europe
2014	Apr 15	Total	N. America, western S. America, Pacific
	Oct 08	Total	W. coast of N. America, Pacific
2015	Apr 04	Total	Pacific, E. Australia
	Sep 28	Total	W. Europe, W. Africa, S. America, eastern N. America
2017	Aug 07	Partial	Australia, Asia, E. Africa
2018	Jan 31	Total	Alaska, Australia, S. E. Asia, Siberia
	Jul 27	Total	India, E. and Central Africa
2019	Jan 21	Total	W. Europe, N. and S. America
	Jan 16	Partial	India, E. Europe, Africa
2020			No partial or total eclipses
2021	May 26	Total	Western N. America, Mid-Pacific
	Nov 19	Partial	N. America, Pacific, Siberia
2022	May 16	Total	Eastern N. America, S. America
	Nov 08	Total	Western N. America, Siberia, Japan
2023	Oct 28	Partial	Asia, Africa, Europe
2024	Sep 18	Partial	W. Europe, eastern N. America, S. America
2025	Mar 14	Total	N. America, western S. America
	Sep 07	Total	Asia, Arabia, east coast of Africa

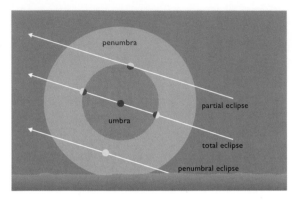

◀ Fig. 7.11 The exact path of the Moon through the Earth's shadow determines the type of lunar eclipse.

▶ Fig. 7.13 The darkness of a total eclipse is affected by the amount of volcanic ash in the atmosphere. The eclipse at left occurred soon after the eruption of Mount Pinatubo.

little interest, because the light is only slightly dimmed and the change is rarely noticeable. Partial eclipses result when part of the Moon passes through the umbral cone, but most attention is given to total eclipses, when it is fully immersed (Fig.7.11). The maximum duration of totality (the period when the Moon is fully eclipsed) comes when the Moon crosses the center of the shadow, and can amount to 1 hr 42 min.

Because the Earth's atmosphere refracts some light into even the center of the umbra, the Moon usually remains visible throughout an eclipse (Fig. 7.12). However, blue light is scattered in the lower atmosphere, so red light predominates and the Moon consequently appears that color. But this is not always the case, and sometimes the Moon has become very dark, or may even have disappeared. The degree of darkness seems to be influenced by several factors, including solar activity, volcanic eruptions, and meteor showers. (Volcanic and meteoric particles can persist in the atmosphere for months, causing increased scattering and darker eclipses.) You can estimate the color and darkness

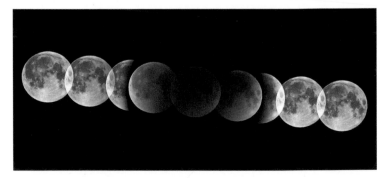

▲ Fig. 7.12 Photographic sequence of a lunar eclipse. At totality, the Moon is only seen by red light, refracted into the shadow by the Earth's atmosphere.

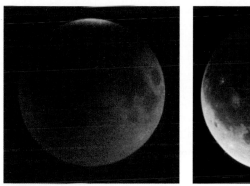

Danjon lunar eclipse scale

0	Very dark, Moon nearly invisible at mid-totality
1	Dark gray or brownish, few details visible
2	Dark- or rust-red with darker central area, outer regions quite bright
3	Brick-red, frequently with a yellowish border
4	Coppery or orange color, very bright with sometimes a bluish border

of an eclipse by using the scale developed by Danjon, and you may find that the value changes during the eclipse.

In a total eclipse, apart from the outlines of the maria, some of the craters may remain visible, especially Aristarchus, Copernicus, Kepler, and Tycho. It has been suggested that this could be (in part) caused by luminescence of some lunar materials, but this seems unlikely. Note down any of these or other features that you are able to see. You might also like to try to obtain a series of drawings during the course of the eclipse, and this is a good opportunity to make some colored sketches.

It is interesting to time the various events during an eclipse. The most obvious are the first and second contacts when the Moon just touches and fully enters the umbra, and the corresponding third and fourth contacts on the way out. In the same way note the times when individual craters enter and leave the umbra. Such timings give information about the way in which the Earth's atmosphere affects the size of the shadow. A very simple observation is to note the time when it first become apparent to the naked eye that an eclipse is happening. This may seem so easy that it can be of little use, but in fact it helps us to judge the accuracy of pre-telescopic observations. They in turn give information about how the Earth's rotation rate has been changing over the centuries.

Lunar and other occultations

An appulse is an apparently close approach between two celestial bodies. Photographs taken on such occasions can be very striking, as are those obtained when planets are near nebulae and clusters. It is well worth keeping an eye on planetary positions with such possibilities in mind.

When the Moon, planets or minor planets pass between the Earth and another object (usually a star), an occultation takes place. Lunar occultations are quite frequent but the motion of the Moon is so complex that many years may elapse before it again occults any particular star. Occultations by other bodies are much rarer, and can usually be seen from only a very restricted area of the Earth, so they are quite a challenge to observers.

For any occultations you need to obtain predictions, and these are published in the yearly handbooks, or are supplied by various national and international organizations. Once again the position of the observer on the Earth makes a lot of difference. Observations provide information about the positions, sizes, and shapes of the occulting bodies, and about the positions and nature of the objects occulted. For the best use to be made of the information, the latitude, longitude, and height above sea level of the observing site should be known as accurately as possible.

The Moon serves as a useful example. As it moves across the sky, stars disappear on the eastern side and reappear in the west – as viewed by the observer (Fig. 7.14). Because of the lack of lunar atmosphere, these events normally happen instantaneously, and may come as quite a shock, especially when they occur at the dark limb. (When you begin this type of work, it is a good idea to observe events in the first part of the lunation, when Earthshine enables the dark limb to be seen, and so provides some warning of the disappearance.) Gradual, or stepped events are sometimes caused by close binary systems (page 180). The bright limb naturally causes considerable interference because of its glare, so

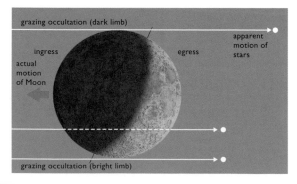

◀ Fig. 7.14 Grazing occultations occur at the northern or southern limbs of the Moon and are less frequent than more central occultations, which can last about an hour.

it is usual to observe disappearances before Full Moon, and reappearances in the later half of the lunation. Bright stars are the exception because both types of event may be seen relatively easily. It is of advantage to use a telescope with a high focal ratio, or at least with a high magnification, so that less of the Moon is in the field of view. The intensity of the light is also reduced, making the star easier to see.

Disappearances present few problems, as the stars may be easily located before the event, but reappearances are a little more difficult. An equatorial mount with setting circles is the best solution, but otherwise the predicted position angle of reappearance will have to be used. On an equatorial mount, a wide-field eyepiece with a crosswire arranged to show the line of drift will enable the point of reappearance to be established, if the point of disappearance has been observed.

Observing occultations may be undertaken with almost any size of instrument. Similarly, the equipment required for recording the times of the events need not be very complicated; ordinary stopwatches, either analog or digital, are frequently used, and various other methods exist. The problems arise in knowing the precise Universal Time (to an accuracy of a fraction of a second) at which events occur. Obviously you cannot use an ordinary clock or digital watch, however accurate these may seem to be for ordinary purposes or other observational work. Any timing equipment must therefore be calibrated against accurate time signals, and these are usually obtained from telephone or short-wave radio services, the latter having the greatest accuracy. Checks are carried out before and after the events so that any changes in the rate of the timing device may be established. Taking all these precautions and observing with care, experienced observers may achieve accuracies of one-tenth of a second. There will be differences between their timings and those of other observers due to the variation in personal reaction times – an effect known as personal equation – but these may usually be taken into account in the full analysis of the observations.

Grazing occultations

Grazing occultations occur when a star appears to just brush the Moon's northern or southern limb. Irregularities in the surface mean that it may then disappear and reappear several times (Fig. 7.15). These events are fascinating to observe, so try not to miss one that occurs in your area. However, they do require even more preparation than normal occultations. There is only a narrow track on the Earth's surface where any particular graze may be seen, so it may be necessary to take a portable telescope to a suitable site. Ideally, several observers should position themselves in a line across the track as then an accurate profile of the Moon may be drawn from the various timings.

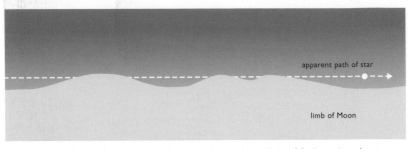

apparent path of star

limb of Moon

▲ *Fig. 7.15 Grazing occultations reveal mountains and valleys on the Moon's* limb, with brief flashes when the star shines through a narrow valley.

Timing is much more difficult, because the star may not only disappear and reappear several times, but also flash out in a narrow lunar valley. The best method is to record both time signals and event markers on the same tape recorder. The "marks" can be spoken words ("In" and "Out," for example), "clicks" made by any suitable means, or better still, a continuous tone produced while the star is invisible. If simultaneous time signals cannot be received by radio, then one has to be obtained by telephone before the event, and the recorder kept running until a second signal has been received after the graze. This is less accurate because temperature changes and declining battery power may alter the recording rate. But with care, it can still give quite good results.

Other occultations

The Moon may also occult planets, and these events are worth observing and trying to photograph. Occultations by planets are also interesting, but here of course, there may be a gradual fade if an atmosphere is present. Observers of Saturn may sometimes see a

star being occulted by the rings. Watch such an event very carefully, recording any changes in brightness that are seen, as well as whether the star is always visible. Once again, timings should be made as accurately as possible. Occultations by minor planets and their results are described later.

◄ *Fig. 7.16 Saturn emerges from occultation at the Moon's bright limb, in this photograph by Paul Stephens.*

THE SUN

Never look at the Sun directly, either with the naked eye, or with any form of equipment. Even the smallest lens concentrates enough light on to the eye to produce lasting damage or complete blindness. This is not surprising when you think how dazzling it is when it happens to shine straight into your face. It may seem a little weaker when low on the horizon, but even then, invisible infrared radiation is still capable of causing damage. It is best to assume that the Sun is always too strong to be observed without special precautions.

The safest, and simplest, method is to project an image. Reflectors are not very suitable for this sort of work, unless they are specially constructed, so use a small refractor. (You can try one side of a pair of binoculars if you have nothing else.) Make sure before you start that any finder (or the second objective of the binoculars) is securely covered by a proper cap. Hold a white card behind the eyepiece, and using the shadow of the telescope as a guide, point it toward the Sun. You can adjust the sharpness of the image by moving the card in and out. Do not point any equipment at the Sun for very long because the concentrated heat could damage the eyepiece, especially if it contains cemented lenses.

Using a card as a projection screen is not very satisfactory, even if a "sunshade" is fitted to the telescope. Construct a lightweight box, with just a small opening for the eyepiece mount, and another so that you can see the image. Try to adjust the size, and the eyepiece used, so that the solar image is a standard diameter – preferably 150 mm (6 inches).

Never think of using ordinary filters to observe the Sun directly – none are safe. This includes even the glass, so-called "Sun" filters occasionally supplied with small refractors. Because these are located near the focus of the objective, they are subject to extreme heating and

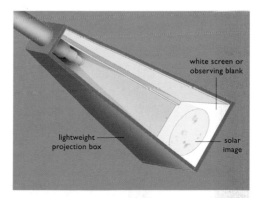

▶ Fig. 8.1 Projecting an image on to a white screen (in this case a projection box) is the only safe way of observing the Sun. In practice, most of the near side would be covered, leaving a small aperture to see the image.

white screen or observing blank

lightweight projection box

solar image

◀ *Fig. 8.2 A small catadioptric telescope with a full-aperture solar filter (here reflecting clouds). Note the additional filter over the finder.*

may easily shatter. Photographic, neutral density, and polarizing filters are particularly unsafe, because without your being able to feel anything, they can transmit harmful amounts of infrared radiation.

Only specially made reflecting filters are safe (Fig. 8.2). These, of metal-coated glass or Mylar film, mount in front of the objective, and allow less than 1% of the radiation to pass. The whole telescope remains cool, which is highly desirable. Even with the small amount of light that is transmitted, it is sometimes necessary to reduce the aperture of the telescope as well. The best filters consist either of two layers of coated material or of a single layer coated on both sides. This prevents any problems that might arise from microscopic pinholes in a single coating.

A proper equatorial mounting is a great convenience. It makes it easier to follow the movement of the Sun, and also helps with orientation. Viewing a projected image may be confusing, but if you move the telescope slightly on each axis in turn, you will soon identify the orientation. With the normal method of projection, the image appears like a naked-eye view but with east and west reversed.

Draw a faint grid of lines on the screen that receives the solar image, and a similar, but heavier grid, exactly the same size, on another piece of card. The second grid will show through a thin drawing blank placed over it, and thus serves to locate the features. When commencing observations, adjust the exact position of the projection box so that either a sunspot or the northern and southern limbs trail along the lines of the grid. (With an altazimuth mount you will have to adjust the orientation at intervals throughout your observing session.)

Solar photography can be carried out quite satisfactorily when a reflecting filter is fitted to the telescope. It is not advisable to try it directly through long-focus lenses without similar precautions, because the concentrated heat may easily damage camera shutters.

Solar features

The Sun is only a fairly small, average star, with a diameter of 1,392,530 km (865,320 miles). From the Earth, at a distance of one astronomical unit, 147,597,870 km (92,960,116 miles), it appears only about 30' in diameter. It is, nevertheless, the only star that we can yet study in detail, and many features can be observed. The apparent surface is known as the photosphere. Its most prominent features are sunspots (Fig. 8.3). These normally consist of a dark center (the umbra) surrounded by a paler, outer region (the penumbra), which under good conditions will often show some radial structure. The smallest sunspots, appearing as tiny dots, are known as pores. A few days' observation shows that spots are carried across the disk by the solar rotation. Because the Sun is completely gaseous, however, the rotation period varies between the equator and the poles, and an average apparent period is about 27.27 days. The apparent paths of sunspots around the disk are influenced by the tilt of the solar axis relative to the Earth. At the limb they usually appear considerably foreshortened.

Sunspots appear dark because they are slightly cooler than the surrounding surface. They are regions where the Sun's magnetic field is particularly strong, and they often form in close pairs of opposite magnetic polarity. Complex spot groups also occur and may cover

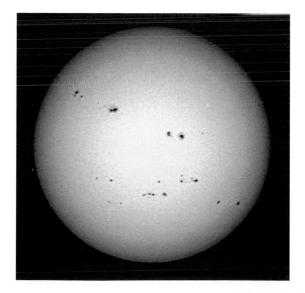

▶ Fig. 8.3 Sunspots are clearly visible on this image of the Sun taken by H. J. P. Arnold. The blue color is a result of the photograph having been taken through a Mylar filter.

considerable areas of the surface. Individual sunspots may be fairly short-lived, appearing and disappearing in just a few days, but groups and centers of activity may be more persistent, and their evolution can be traced by daily observation during the 7–10 days that they are easily visible on the disk. They may even reappear over the limb after a complete solar rotation. Counting active areas is a relatively simple, but very useful observational task.

Overall numbers of spots fluctuate in the 11-year sunspot cycle. The cycle actually affects general solar activity, of which sunspots are only a small, easily visible part. The course of an individual cycle begins at minimum when small spots make their appearance at high solar latitudes in each hemisphere, although rarely above 35°. The general centers of activity migrate toward the equator, and at sunspot maximum are concentrated around latitude 15°. After this the number declines, but while the old areas of activity continue to move toward the equator, the first spots of the next cycle begin to form at high latitudes.

Bright patches are known as faculae, and exist both before and after sunspots form in the same areas. They are most visible toward the solar limb, unlike the much smaller granulation, which appears under good conditions on the center of the disk. Granulation consists of cells with light centers and darker borders, and thus produces a generally mottled appearance. The edges of the Sun always show limb darkening, where we are viewing cooler, higher regions than in the center of the disk.

Certain experienced amateurs use specialized equipment to view the Sun at a single wavelength of the spectrum. The most common method is to employ special filters (known as interference filters) to remove all light except for a narrow band at a particular wavelength, usually "H-α," a hydrogen emission line. Observation at this wavelength reveals certain individual features on the Sun, most particularly prominences. These are glowing clouds of gas that project into space beyond the limb of the Sun; they appear as dark streamers (known as filaments) when seen against the glowing disk.

Solar eclipses

When the Moon passes between the Sun and the Earth at New Moon, the three bodies are rarely perfectly aligned, and the Moon's shadow usually misses the Earth. However, at least twice a year, and sometimes as many as five times, the shadow does touch the Earth and produces a solar eclipse.

As with the shadow of the Earth, the Moon's consists of a dark, central umbra and a larger, outer penumbra. When just the latter touches the Earth, observers see a partial solar eclipse, with only some of the disk being covered (Fig. 8.4). Partial eclipses are not of very great interest to astronomers, although they do give some opportunity for photography. Do please remember that just the same precautions must be followed when observing partial eclipses as with observing the Sun itself. Never look at them directly with the naked eye, through binoculars or a telescope, because damage to your eyesight will almost certainly result. Proper solar filters must be used for visual observation, and either a solar filter or a very dark neutral-density filter (with the unusually high density factor of 5.0) must be employed if you try any photography with conventional camera lenses.

If the umbra reaches the Earth, a total eclipse is produced (Fig. 8.4). Bright stars and planets may become visible in the darkened sky. The zone of totality is only small (no more than about 300 km, or 190 miles, across at the most), and it is for this reason that total solar eclipses are only rarely seen from any particular place on Earth (Fig. 8.5). The path of totality sweeps across the surface as a result of the combined effects of the Earth's rotation and the motion of the bodies. The maximum duration of the total phase is about 7 minutes 30 seconds, but because of the varying distance of the Moon it can happen that the umbra only

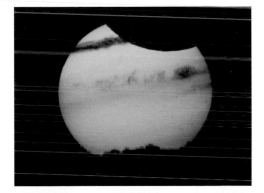

▶ Fig. 8.4 The various types of solar eclipse: a partial eclipse (right), slightly obscured by clouds; an annular eclipse (below right); and a total eclipse (below).

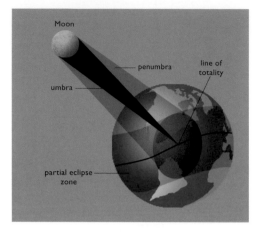

◀ Fig. 8.5 As the elongated cone of the Moon's shadow sweeps across the Earth, a total eclipse will only be seen in a narrow central track.

"touches down" at one point for a few seconds. At some eclipses the umbral shadow cone fails to reach the ground, giving rise to an annular eclipse. The Sun will not be completely covered but will appear as a ring surrounding the Moon. Although not as fascinating as a total eclipse, an annular eclipse is still a dramatic sight (Fig. 8.4).

As with lunar eclipses, significant times are known as "contacts." We have:

First contact	Moon's disk just touches Sun: penumbra reaches observer and partial eclipse begins
Second contact	Moon's disk completely within Sun's disk: umbra reaches observer and total or annular eclipse begins
Third contact	Moon's disk touches opposite side of Sun's disk: umbra leaves observer and total or annular eclipse ends
Fourth contact	Moon's disk finally uncovers Sun's disk: penumbra leaves observer and partial eclipse ends

Forthcoming solar eclipses are listed in Table 8.1. In this table, the durations quoted are those of the total or annular phases, that is, from second to third contact.

The most conspicuous solar feature revealed at a total eclipse is the corona, which spreads far into space. It is the outermost layer of the solar atmosphere, and has a temperature of 1–2 million degrees C. Its shape and size change with the sunspot cycle, being more regular at sunspot maximum. At sunspot minimum it is more asymmetrical, with long equatorial "streamers" as well as shorter "plumes" at the poles. The inner corona may be studied at any time with special equipment, but the outer region is seen only at eclipses.

Equally striking are the bright, pinkish prominences seen around the limb. These loops and wisps of glowing gas may appear to be material ejected from the surface, but are often gas streaming down from the corona. When the Moon almost covers the solar disk, brilliant points of light may remain visible through lunar valleys in the effect known as Baily's Beads. At the very beginning and end of totality a single uncovered part of the photosphere may give rise to a striking "Diamond Ring" effect.

When the Sun is fully eclipsed, the planets and the brighter stars frequently become visible. More significantly, comets that were previously undetected (because their paths lay close to the Sun in the sky) may be revealed. Discoveries of such "eclipse comets" are now

TABLE 8.1: FORTHCOMING SOLAR ECLIPSES				
Date		Type	Duration	Area of visibility
2012	May 20	Annular	5m 46s	S. Japan, N. Pacific, W. coast of N. America
	Nov 13	Total	4m 2s	Queensland, S. Pacific
2013	May 10	Annular	6m 4s	N. Australia, S. Pacific
	Nov 03	Annular/total	1m 40s	N. Atlantic, Equatorial Africa
2014	Apr 29	Annular	0m 0s	S. Indian Ocean, Australia
	Oct 23	Partial		Western N. America
2015	Mar 20	Total	2m 47s	N. Atlantic, Arctic Ocean
	Sep 13	Partial		S. Africa, S. Indian Ocean, Antarctica
2016	Mar 09	Total	4m 10s	Indonesia, N. Pacific
	Sep 01	Annular	3m 6s	Central Africa, Indian Ocean
2017	Feb 26	Annular	0m 44s	S. America, S. Atlantic, Central Africa
	Aug 21	Total	2m 40s	N. America
2018	Feb 15	Partial		Antarctica
	Jul 13	Partial		Southern Ocean
	Aug 11	Partial		Siberia, Canadian Arctic
2019	Jan 06	Partial		N. Pacific, Japan, Siberia
	Jul 02	Total	4m 33s	S. America, Southern Pacific
	Dec 26	Annular	3m 39s	N. Australia, Asia, India
2020	Jun 21	Annular	0m 38s	Asia, E. Africa, E. Europe
	Dec 14	Total	2m 10s	S. Atlantic, S. America, Pacific
2021	Jun 10	Annular	3m 51s	Arctic, Arctic Canada
	Dec 04	Total	1m 54s	S. Atlantic, Antarctica
2022	Apr 30	Partial		S. E. Pacific
	Oct 25	Partial		W. Asia, Europe, N. E. Africa
2023	Apr 20	Annular/total	1m 16s	W. Pacific, Indonesia, S. Indian Ocean
	Oct 14	Annular	5m 17s	Northern S. America, C. America, S. & W. USA
2024	Apr 08	Total	4m 28s	Eastern N. America, Mexico, Pacific
	Oct 02	Annular	7m 25s	S. E. to N. W. Pacific

less common because space probes, such as the SOHO mission, now monitor the Sun continuously, and detect many comets that would previously have gone unnoticed.

Observing eclipses

Dedicated "eclipse chasers" often travel to the ends of the Earth, and go to extreme lengths – such as hiring an aircraft to get a better view, or racing across country to find a break in the clouds – to observe every solar eclipse. Because eclipses may occur in extremely remote places, following such events can take on the nature of a full expedition. Even for more accessible eclipses, transporting equipment can pose considerable problems.

Because eclipses are generally short, considerable experience and preliminary practice are required to carry out extensive photography or any scientific work. If it is your first eclipse, it is probably best not to try anything ambitious – such as extensive photography – but to aim to enjoy the spectacle. It may be practical to take some photographs of (say) the surrounding countryside as the Moon's shadow sweeps across the Earth. Photography of the different objects visible during the total phases (corona, prominences, Baily's Beads, Diamond Ring, and so on) is complex, involving considerations of equipment, focal lengths, filtration, film speeds, and other variable factors, so it is not discussed here. Instead, the reader is referred to the books listed for further reading. If you are part of a group, it is often possible to divide the activities between observers to allow everyone some time simply to look at the sky with the naked eye and enjoy the spectacle.

During the totality it is perfectly feasible to use binoculars or other instruments to sweep for planets or comets, or even examine prominences around the limb of the Sun. Because time appears to pass extremely rapidly during an eclipse, great care needs to be taken to ensure that you are aware of the end of totality. Some observers use a small timer to warn them of the approach of third contact, when filters and lens caps may be refitted, binoculars put aside, and full safety precautions reinstated.

There are, of course, a number of phenomena that do not directly involve the Sun and Moon. In addition to the way in which the Moon's shadow may be seen racing across the Earth, there are the changes in lighting and the color of the sky. At very narrow crescent phases, rippling shadows, known as shadow bands, may be seen passing across light-colored objects such as walls and the ground. Depending on the observing site, the behavior of animals and birds may also be particularly notable.

— OBSERVING THE PLANETS —

If you have a choice, use a refractor, Cassegrain reflector or catadioptric telescope for planetary observations. Their high focal ratios give larger primary images than Newtonian reflectors, and the restricted fields are no disadvantage. If you already have a Newtonian, you can still get excellent results if you use a Barlow lens and eyepieces of good quality. A 75 mm refractor is the minimum size for seeing any planetary detail, but as always, large apertures are of advantage with their greater light-grasp – which itself allows a higher magnification to be used – and their finer resolution.

The changes that occur on most of the planets mean that you never know what to expect when you go to the telescope. When you start observing do not be too disappointed by the tiny disks and the fact that you seem to see very little – it takes a while for your eyes to "learn" to make out faint details. As you get used to your equipment and have some practice, you will find that you see more and more. You will also find that there are tantalizing occasions when the seeing becomes perfect – usually for only too short a time – and the tiny disks are covered in so much detail that you have a difficult task in capturing the appearance in a drawing. Do not forget, too, that it may be just as important to know that no details were visible, as can sometimes happen, especially in the cases of Venus and Mars, so always make a note of these "negative" observations.

It does no harm just to look at the planets casually from time to time, and the more frequently you do this the more you are likely to see. Trying to make proper observations is even more satisfying, and it certainly helps you to become familiar with the planet's appearance. This is important because once you have gained that experience, anything unusual that happens will be immediately obvious.

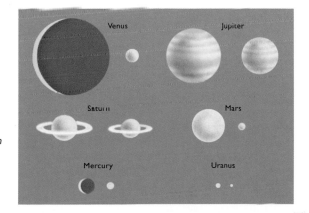

► Fig. 9.1 The maximum and minimum apparent sizes of the planets. Jupiter shows the least relative change.

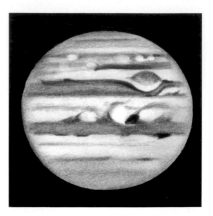

▲ Fig. 9.2 Drawing of Saturn
(south at top) by Richard McKim.

▶ Fig. 9.3 Jupiter (south at top)
drawn by Richard McKim.

Actual observations that you can carry out are: making whole-disk or detail drawings; estimating intensities and phases; and timing transits of features across the central meridians of the disks. (Photographic work tends to be so specialized that it is not considered here.) If you are a beginner you will probably want to start with whole-disk and detail drawings. Jupiter and Saturn offer the added attraction of various satellite phenomena, and these are discussed in the next chapter.

One point that deserves to be mentioned is that confusion can arise over the the terms "east" and "west" when they are used to refer to planetary (and lunar) features. Before exploration by spacecraft became common, the usage was always applied in the same way as sky orientations, so that in the naked-eye view, a feature was east of another if it was to the left (as seen by a northern-hemisphere observer). However, with highly detailed spacecraft mapping (and to prevent confusion for astronauts on the Moon), it was obviously sensible for latitude and longitude to apply in the same way as on the Earth. This reversed the two directions east and west, so that, for example, Mare Orientale ("Eastern Sea") is now west of the central meridian of the Moon. Planetary observers frequently avoid problems by using the terms "preceding" and "following" to describe the positions of features.

Like the Moon, the planets have terminators dividing the illuminated from the unilluminated portions. In the cases of Mercury and Venus the terminators and the resulting phases are easily visible. With Mars it is more difficult to detect the phase, although it is often present, and Jupiter and Saturn are so distant from the Earth and the Sun that the terminators are, to all intents and purposes, the same as the visible limbs. The rotation of the planets gives rise to morning and evening terminators, and in some cases this may be related to the appearance or occurrence of particular features (especially Martian clouds).

Planetary drawings

Before you can make planetary drawings you need some suitable blank outlines. Even though the apparent planetary sizes differ greatly around their orbits (Fig. 9.1), particularly in the cases of Mars and Venus, most organizations that coordinate amateur work use fixed diameters for particular planets so that observations may be easily compared. Try to keep to these sizes if you draw any blanks for yourself.

There are some other points that have to be borne in mind. Both Jupiter and Saturn are considerably flattened by their rapid rotation, so their outlines are not perfectly circular, and you cannot use a pair of compasses to draw a blank. Their blanks must be specially prepared, by using a template or tracing from a proper outline (pages 157 and 162). Saturn, of course, has the added complication of the changing aspect of the rings (Fig. 9.2). Observational groups issue a whole series of blanks to take account of the differences in the appearance of the rings. Mercury, Venus, and Mars can be drawn with circular outlines, which helps, but they show phases, somewhat like the Moon's, where part of the hemisphere turned toward Earth is not illuminated. With Mercury and usually in the case of Mars, the amount of phase can be accurately predicted, so you can draw it in advance. However, this does not apply to Venus, and the terminator must be added from actual observation. The amount of phase effect for Jupiter and Saturn is so small that it may be ignored.

Just as you need some experience to be able to see the detail, so you need practice to be able to show it in drawings. This is particularly the case with Mars, Jupiter, and Saturn because of the amount of detail that they may show, and also because their rapid rotation means that

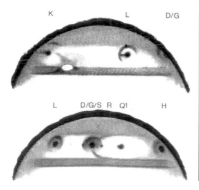

▲ Fig. 9.4 McKim's drawings of some of Comet SL-9's impact sites on Jupiter.

▲ Fig. 9.5 Whole-disk drawing of Mars by Richard McKim.

the appearance changes during an observing session. You might find it better to start by trying to reproduce just the overall distribution of light and dark areas, or concentrating on some specific feature (Fig. 9.4), rather than attempting fully detailed, whole-disk drawings (Fig. 9.5). As you gain experience, you can add more detail. In any case, as has been said before, it is probably as well to begin every drawing in this way.

Intensity estimates

Generalized drawings can help to lead on to making intensity estimates. In these, you assess particular features in terms of their relative brightness, and give them numerical values. It is not as difficult as it sounds. The scale depends upon the planet and the range of brightness that it shows, but normally a value of 0 corresponds to white, and the numbers increase for darker features, with 10 representing the black sky background. Some organizations use a reversed scale (that is, one of brightness, with 10 implying a brilliant white feature and 0 implying one that is absolutely black). Unfortunately, but not surprisingly, there are usually differences between the values that various observers give to the same feature. This also partly depends upon the equipment that is used, including the magnification, as well as the seeing conditions. There may be additional confusion over the exact identifications, so it is a good idea to show intensity estimates on an actual drawing of the planet, even if it is only a rough sketch (Fig. 9.6).

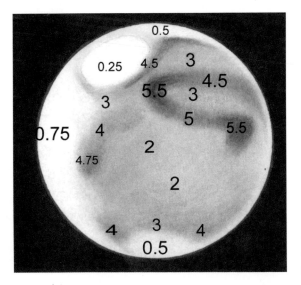

◀ Fig. 9.6 An intensity diagram of Mars by Richard McKim.

► *Fig. 9.7 Images of Mars taken by Don Parker, without filters (top), and through red, green, and blue filters, each accentuating different features.*

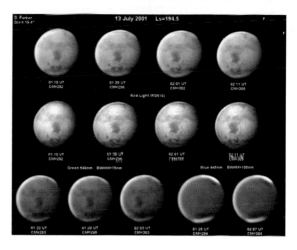

Filters

You might like to try using various filters to help make ordinary drawings or intensity estimates. However, unless your telescope is fairly large, they may do more harm than good, because of the inevitable light losses produced. Venus is the exception, as it may be so bright that a neutral-density filter – or daytime observing – is useful in diminishing the glare, and thus makes the details more easily visible. Generally, Mars and Venus are the most satisfactory subjects for color-filter observations, although there is no reason why they should not be used on Jupiter and Saturn. On Mars, for example, a light blue filter will accentuate atmospheric features, whereas one of an orange or reddish tint might show greater surface detail (Fig. 9.7). Some eyepieces are specially threaded so that optical glass filters can be screwed into place. These are ideal, but tend to be expensive. Ordinary photographic gelatine filters cost much less and you can cut and mount them easily, either in simple holders or a special adaptor like that used for lunar filter observations. They are more delicate, however, and tricky to clean. Never mount any filters close to the objective's focal plane where any defects will be in focus and glaringly obvious. Photographs taken through filters of various colors make an interesting experiment.

Transit timings

As a planet rotates, the various features are carried across the central meridian of the disk (Fig. 9.8). The timing of these central meridian transits is a very valuable technique, even though it may not sound very appealing. You can use the times (accurate to about a minute) to find the actual longitudes of individual markings on the planets. Tables, given in the various handbooks, usually for both Mars and Jupiter,

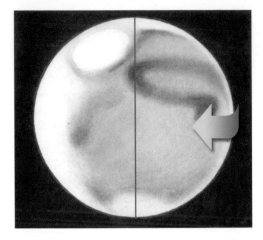

◀ *Fig. 9.8 Either a real (crosswire) or imaginary central meridian may be used for planetary transit timings.*

show the longitude on the central meridian at 00:00 UT, and how much it changes with particular intervals of time. From these you can easily establish the longitude of any feature that you observe. If you observe markings on more than one occasion, the longitudes can be compared to see whether there has been any movement during the interval. This is a very accurate method of recording the positions of planetary features. It is particularly fascinating to see the changes that take place on Jupiter, as some spots wander about and overtake others.

Another type of transit, that of the inferior planets across the Sun, is discussed on page 151.

The inferior planets

The two planets inside the orbit of the Earth (the inferior planets) are Mercury and Venus. Because of their orbits, they show a full range of phases from the thinnest crescents, when they are at inferior conjunction between the Earth and the Sun, to "full" phases at superior conjunction on the far side of the Sun. Because they are always close to the Sun, they never appear far above the horizon if you observe before sunrise or after sunset. For this reason observations are often made in actual daylight, usually in the period of about two hours after sunrise or before sunset, when the conditions are reasonably steady. In the case of Venus the reduced contrast between the planet and the background sky also means that more detail can then be seen.

Mercury

Mercury can reach about magnitude -1.7, not very different to Sirius (-1.4), but it is never more than about 28° from the Sun, so the first problem comes in locating it. If you live in the northern hemisphere it is easiest to find the planet at eastern elongations in the spring, and western elongations in the autumn, when it will appear highest, although its actual distance from the Sun under these conditions is never more than 18°. Greater elongations only occur when the planet is low in the sky. Conditions are best if you live in the southern hemisphere, because the planet can be both higher in the sky, and at its greatest elongation of about 28° in the morning in autumn (April), and in the evening in spring (September) – at western and eastern elongations respectively. Observers in the tropics are generally well placed to observe the planet at any elongation. Forthcoming elongations are given in Table 9.1, with distances from the Sun rounded to the nearest degree.

The only really satisfactory and infallible way of finding Mercury is by using setting circles, but because it is so close to the Sun, always take great care, for safety's sake. When the planet is at eastern elongation, you can move so that the Sun is hidden behind a building. This is

TABLE 9.1: ELONGATIONS OF MERCURY, 2012–2022							
	Eastern elongations				Western elongations		
2012	Mar 05	18°	Jul 01	26°	Apr 18	28°	Aug 16	19°
	Oct 26	24°			Dec 04	21°		
2013	Feb 16	18°	Jun 12	24°	Mar 31	28°	Jul 30	20°
	Oct 09	25°			Nov 18	19°		
2014	Jan 31	18°	May 25	23°	Mar 14	28°	Jul 12	21°
	Sep 21	26°			Nov 01	19°		
2015	Jan 14	19°	May 07	21°	Feb 24	27°	Jun 24	22°
	Sep 04	27°	Dec 29	20°	Oct 16	18°		
2016	Apr 18	20°	Aug 16	27°	Feb 07	26°	Jun 05	24°
	Dec 11	21°			Sep 28	18°		
2017	Apr 01	19°	Nov 24	22°	Jan 19	24°	Apr 01	19°
	Jul 30	27°						
2018	Mar 15	18°	Jul 12	26°	Jan 01	23°	Apr 29	27°
	Nov 06	23°			Aug 26	18°	Dec 15	21°
2019	Feb 27	18°	Jun 23	25°	Apr 11	27°	Aug 09	19°
	Oct 20	25°			Nov 28	20°		
2020	Feb 10	18°	Jun 04	24°	Mar 24	28°	Jul 22	20°
	Oct 01	26°			Nov 10	19°		
2021	Jan 24	19°	May 17	22°	Mar 06	27°	Jul 04	21°
	Sep 14	27°			Oct 25	18°		
2022	Jan 07	19°	Apr 29	21°	Feb 16	26°	Jun 16	23°
	Aug 27	27°	Dec 21	20°	Oct 08	18°		

◀ *Fig. 9.9 Mercury and the waning Moon, 1999 August, photographed by Tunç Tezel.*

the only time when you can safely sweep for the planet, because the apparent motion of the Sun only puts you farther into shadow. Obviously the technique would not be safe for western elongations, but it is then easiest to find the object before sunrise. You can keep track of the planet as it moves into the daylight sky ahead of the Sun.

Unfortunately, once Mercury has been located there is little that can be observed (Fig. 9.9). Its color is often dull white, and this is most apparent when the more brilliant Venus is nearby in the sky, but it may also have a red or yellowish tinge. You will see the phases in a small telescope of about 75 mm aperture, but that is the most that is likely to be visible. Observers have distinguished some of the darkest markings with apertures of only 100–150 mm, but do not expect to see very much even with considerably larger telescopes. It has been suggested that the details are somewhat easier to see if a pale yellow filter is used, and this is worth trying.

Venus

Venus is much more satisfactory to observe, and being brighter than Mercury (it may reach magnitude −4.3), it is usually much easier to find. Its elongation may reach 47°, when it is accessible for a few weeks. Forthcoming elongations and inferior conjunctions are given in Table 9.2. This table is arranged to show the sequence as the planet moves from eastern elongation (in the evening sky), through inferior conjunction (when it is largest and generally closest to Earth), through western elongation (in the morning sky), to superior conjunction (when it is on the opposite side of the Sun).

In certain years the planet may be visible for months at a time. You will find that observations in daylight are the most satisfactory, because the glare is considerably reduced, allowing the faint details to be seen. If you use a telescope you may still need a neutral-density filter to diminish the amount of light from the planet. Venus is much

TABLE 9.2: ELONGATIONS AND CONJUNCTIONS OF VENUS, 2012–2024							
Eastern elongation			Inferior conjunction		Western elongation		Superior conjunction
2012	Mar 27	46°	2012 Jun 06		2012 Aug 15	46°	2013 Mar 28
2013	Nov 01	47°	2014 Jan 11		2014 Mar 22	47°	2014 Oct 25
2015	Jun 06	45°	2015 Aug 15		2015 Oct 26	46°	2016 Jun 06
2017	Jan 12	47°	2017 Mar 25		2017 Jun 03	46°	2018 Jan 09
2018	Aug 17	46°	2018 Oct 26		2019 Jan 06	47°	2019 Aug 14
2020	Mar 14	46°	2020 Jun 03		2020 Aug 13	46°	2021 Mar 26
2021	Oct 09	47°	2022 Jan 09		2022 Mar 20	47°	2022 Oct 22
2023	Jun 04	45°	2023 Aug 13		2023 Oct 23	46°	2024 Jun 04

larger than Mercury, and comes closer to the Earth, so you can even see the phases with good binoculars. To make out any details you still need apertures of at least 100 mm, and preferably more.

The only details visible on Venus are those of the uppermost layer of clouds in the dense atmosphere, so they are, at best, ill-defined and rather difficult to draw. However, it should be remembered that amateur observations of these faint markings obtained a rotational period for the upper atmosphere of about four days well before this was confirmed by spacecraft measurements. (The rotation of the invisible, solid body of Venus is retrograde, and has a period of 243 days.)

You may see both light and dark markings on the disk, but it is frequently difficult to show these in any drawing without exaggerating

▲ Fig. 9.10 The motion of Venus in the sky. The sequence runs from 2000 September (bottom center), through 2000 December (left) to 2001 March (right). The photographic sequence was obtained by Tunç Tezel.

▲ Fig. 9.11 This drawing of Venus shows the bright cusps that are sometimes seen on the planet.

► Fig. 9.12 This photograph by Tunç Tezel of a narrow crescent Moon and Venus distinctly shows the planet's phase.

the contrast (Fig. 9.11). Make a note if you find that this is necessary. If possible, make proper intensity estimates at the same time. Bright "cusp-caps" are frequently recorded, generally, but not always precisely, over the polar regions. Apparently darker "collars" around these caps are sometimes visible. A filter can help to accentuate the details on Venus and a light yellow (Wratten 15) is the best to try. As always, however, it is still important to make a note when there is no visible detail.

The horns of the crescent may sometimes appear unequal, being either blunted or extended. The terminator can also seem irregular, rather than a smooth curve. Try to record these changes in a careful drawing. The irregularities on the terminator may make it more difficult to determine the planet's apparent phase, which generally differs from the one predicted. In particular, at half-phase, or dichotomy, the discrepancy may amount to several days. (This is known as the Schröter Effect, after the observer who first noted it.) Dichotomy is early at eastern elongations, and late at western ones. Although this effect is definitely real, no cause has yet been established. Because the terminator position is not easy to record, several drawings are really required from each observing session, so that an average value of the observed phase can be deduced. A micrometer is the most accurate method of obtaining these results, but unfortunately very few observers have one.

Another uncertain effect is that of the "Ashen Light." When the crescent is very narrow, the dark portion of the disk may appear to be faintly luminous, similar to the effect of Earthshine on the Moon. The only chance of seeing this effect is if you can fit an eyepiece with a home-made, occulting bar, shaped to hide the bright crescent. Although it has been suggested that perhaps some auroral phenomenon might be involved, it is still possible that it is just an optical illusion, like the opposite effect where the unilluminated side of Venus appears to be darker than the surrounding sky.

Transits

Mercury and Venus occasionally cross the disk of the Sun. These events, known as transits, must be observed with safe methods like those used for studying solar features. Transits of Mercury are more frequent than those of Venus and often occur in pairs with three years between them. They can occur only in May or November. Venus transits occur in pairs (separated by eight years) with more than a century between each pair. Again, they occur only in June or December. The latest pair of transits took place in December 1874 and December 1882; the first of the 21st-century pairing was in June 2004 with the next transit due in June 2012; the subsequent pair will be in December 2117 and December 2125. Although transits were once of great scientific interest, their observation nowadays is purely for the pleasure of seeing a rare phenomenon. A series of drawings or photographs of the solar disk showing the motion of the planet relative to any sunspots forms a striking record of such an event.

Recent and forthcoming transits are listed in Table 9.3. The duration of transits depends on whether the planet appears to cross the Sun's

TABLE 9.3: TRANSITS OF MERCURY AND VENUS				
Mercury		Start	Center	End
	1993 Nov 06	03:07	03:58	04:48
	1999 Nov 15	21:16	21:42	22:08
	2003 May 07	05:14	07:53	10:33
	2006 Nov 08	19:13	21:42	00:11
	2016 May 09	11:13	14:59	18:44
	2019 Nov 11	12:37	15:21	18:05
Venus		Start	Center	End
	1874 Dec 09	01:49	04:07	06:26
	1882 Dec 06	13:56	17:06	20:15
	2004 Jun 08	05:15	08:21	11:27
	2012 Jun 06	22:11	01:31	04:51
	2117 Dec 11	00:02	02:52	05:42
	2125 Dec 08	13:19	16:06	18:52

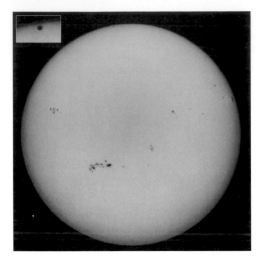

◀ Fig. 9.13 Mercury can just be seen (top and inset) during its transit of the Sun on 1992 November 15.

disk in the polar or equatorial regions. (For complex reasons that cannot be discussed here, the times given differ very slightly from Universal Time, but are sufficiently accurate for most purposes.)

Mars

Mars is a wonderful object to study, so it is a great pity that it is not favorably placed every year. Oppositions are about 780 days apart (Table 9.4 and Fig. 9.14), and when these occur near aphelion (in January and February) the apparent diameter may be as small as 13″. It can rise to nearly 26″ at oppositions near perihelion (in August and September). Anyone living in the southern hemisphere has a wonderful opportunity at these oppositions, because Mars is then south of the ecliptic, and very low for observers north of the equator (as shown by the declinations given in the table). The south pole of Mars is turned toward the Earth on these occasions, the north pole being visible only when Mars is around aphelion. Despite the difficulties, good observers can still see a lot of detail at most oppositions.

There are numerous dark markings on the lighter background of the Martian disk. As the planet rotates and different longitudes come into view, new features slowly become visible from night to night. With patience, a map can be built up, showing the appearance right round the planet. Most of the markings are definitely permanent, and are seen at every apparition. Other markings show changes when observed over a period of years. For a long time it was thought that these alterations might be due to vegetation, but we now know that the winds move material from one area, depositing it in another. Dust

TABLE 9.4: OPPOSITIONS OF MARS, 2012–2022			
Date	Declination	Diameter	Magnitude
2012 Mar 03	+10° 17′	14″	−1.0
2014 Apr 08	−05° 08′	15″	−1.3
2016 May 22	−21° 39′	19″	−1.8
2018 Jul 27	−25° 30′	25″	−2.6
2020 Oct 13	+05° 27′	23″	−2.4
2022 Dec 08	+25° 00′	17″	−1.9

storms sometimes completely obliterate the dark markings over the whole of the planetary surface, particularly at perihelic oppositions. It is then interesting to watch the features, perhaps changed in outline or intensity, gradually reappear as the veil of dust subsides. Lesser storms may affect individual regions of the surface. These dust storms are the "yellow hazes" sometimes mentioned by earlier observers.

The brilliant polar caps wax and wane with the Martian seasons, and the southern cap may even disappear completely. Portions may become detached before they gradually dwindle away, and other rifts are sometimes visible. The darker collars surrounding the caps do not appear to be entirely due to contrast effects – changes in the covering of dust are probably involved. Frequently, however, the features in the polar regions are masked by a "polar hood" of cloud, which sometimes extends over 50–60° of latitude. This cloud tends to disperse at midwinter, when it is frozen out on to the surface, reappearing in the spring as the warmth of the Sun increases and turns the ice back into vapor. We now know that the permanent northern cap is ordinary

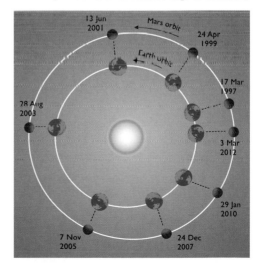

▶ Fig. 9.14 Some years are more favorable than others for observing Mars. The opposition in 2003 August was the closest for 60,000 years.

water ice, but the seasonal caps consist of both water and carbon dioxide. The latter only freezes during the coldest part of the winter.

Other whitish "hazes" are sometimes visible elsewhere on Mars, most particularly at the morning terminator, where clouds formed during the night have not yet dispersed. But they are not confined to just this region, and some can be followed as they move across the surface. Other "blue hazes" may cover very considerable areas of the disk.

Observing Mars

Once again you really need an aperture of at least 100 mm to be able to see any proper details of the surface, and 150 mm would be better for regular observation. Magnifications of 200–400 are likely to be the most satisfactory.

The positions of the major dark features are fairly well established, but good drawings of individual markings are always of interest. However, try to make at least one or two whole-disk drawings at each apparition. Despite the considerable changes in the diameter of the planet it is probably best to use a single size of observing blank, 50 mm (2 inches) in diameter, all the time. The phase may amount to as much as 46° and so needs to be accurately reproduced. You can find details of the phase at any time in one of the astronomical yearbooks.

Because Mars rotates fairly rapidly – its rotation period is $24^h 37^m$, not much greater than that of the Earth – you should not take too long in making a detailed, whole-disk drawing. Start by locating the polar cap, but remember that this may not be centered precisely on the pole. (The position of the poles depends upon the exact tilt of the axis as seen from Earth, but once again, a yearbook will give you that information.) Next, sketch the most prominent dark features, and make a note of the time when you finish this basic drawing. You can then carry on, adding the finer details and outlining the brighter areas.

A light blue filter is very useful for showing atmospheric features, and a Wratten 44B is recommended. Although other, strongly colored filters have often been suggested in the past, they are only suitable for experienced observers using large instruments. In most small telescopes, Mars does not appear highly colored, because of the limited light-grasp, and dense filters will only degrade the image, rather than enhance it.

Intensity estimates

Intensity estimates are very valuable, and should be attempted. Like many other astronomical observing techniques, they are not that difficult after a little practice. Use a scale running from 0 (brightest) to 10. The polar caps are usually taken as intensity 0, and a black sky background as 10. The latter in particular, however, does depend upon the equipment and magnification being used. The sketch accompanying the intensity estimates may be quite rough and "unfinished" provided the features are easily identifiable (Fig. 9.6, page 144).

Transit timings

You can also try making central meridian timings. A little practice soon enables you to judge when a feature is exactly halfway across the disk. Do not try to make lots of timings of very faint details, as these may be difficult to identify. The bolder, more distinct features are

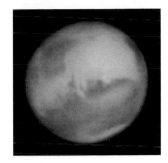

▲ Fig. 9.15 A video image of Mars by Steve Massey.

far better. In any case, timings can only be carried out when Mars appears "full." This is only for about 7–10 days each side of opposition. At other times the phase makes the task impossible.

Photography

Photography of Mars really requires large apertures and long focal lengths to give reasonably sized, bright images. In recent years, some amateurs with large telescopes have used CCD cameras to obtain some remarkable images of Mars. Only a few years ago such pictures were difficult to obtain even with professional-size telescopes. Anyone using conventional film, however, requires perseverance, and needs to make the most of occasions of exceptionally good seeing. The best such results have usually been with black-and-white films, sometimes exposed through different filters, although some success has been achieved with color films.

Minor planets

Most amateurs find that the main challenge in observing minor planets (or asteroids as they are frequently called) comes in locating and following them. Most of the orbits lie between those of Mars and Jupiter, and they are all quite small – the three largest being Ceres, diameter 1000 km (620 miles), and Pallas and Vesta, both with diameters of about 540 km (330 miles). They are faint; Vesta is the only one that can rise to just about the naked-eye limit and perhaps half-a-dozen others may exceed magnitude 10 at opposition. Nevertheless, there is a considerable sense of achievement in managing to track one down from the details given in the yearly handbooks, and even more in keeping it in view for a period of time. The best way of doing this is undoubtedly to plot the positions on charts which show stars fainter than the objects' expected magnitudes.

Photography

Photography can be attempted with any driven equipment, and depending upon the minor planet's position in its orbit should show the motion on exposures taken on different nights. Obviously this may not be the case if it is near one of its stationary points. Although there is no reason why 35 mm cameras should not be used, better results will generally be obtained with those that use larger film sizes, while still giving a fairly wide field of view. The objects that come close to the Earth, or cross its orbit, may sometimes move so rapidly that they can be recorded as trails. Good-quality photographs can be used to obtain positions, and thus refine the orbits of some of the poorly known minor planets, but these generally require special attention to equipment and methods. Here again, experienced amateurs using CCD equipment have achieved considerable success.

Some minor planets show changes in brightness, because they are irregularly shaped and rotating. The magnitudes can be estimated by the methods generally used for variable stars, but unfortunately it is usually very difficult to obtain satisfactory magnitudes of comparison stars. Photographs of the surrounding star fields taken through the appropriate filters can be of help here, but most progress is likely to be made by those dedicated amateur astronomers who have CCD equipment.

Occultations

One fascinating field is that of occultations, predictions for which have become possible in recent years. The methods used are like those for lunar grazing occultations, and similar occultation tracks can be prepared. The information gained from accurate timing gives precise measurements of the sizes of minor planets, and in some cases their individual shapes. Possible satellite bodies have also been recorded.

If the star is brighter than the minor planet, it is of advantage to use a small telescope so that the latter remains invisible. The star will suddenly vanish and reappear, and there will be no confusion from the two images merging as may happen when a larger aperture is used.

There are considerable errors in the orbital information for many objects and also in many stellar positions, so last-minute checks must be made by the professional astronomers engaged in this work to refine the predictions. A change of hundreds of miles may be produced by only small errors. Amateurs are frequently able to move to another observing site to compensate for this. Refined predictions of the tracks are now made available over the Internet.

—— OBSERVING THE OUTER PLANETS ——

Jupiter

Jupiter is probably the most fascinating of the planets to observe. It is one of the four "gas giants" – Saturn, Uranus, and Neptune are the others – and mainly consists of the light elements hydrogen and helium. Jupiter's visible markings occur in the uppermost layers of its deep atmosphere, which contains many other gases, including methane and ammonia. There are both large-scale markings, and many smaller features that are always changing. If you only have a small telescope of about 50 mm in diameter, you will still be able to see that the disk is divided into dark belts and polar regions, and brighter zones (Fig. 10.1). But even these major features are by no means permanent, as they strengthen and fade, and divide into more than one component. Apertures of 150 mm or more are really needed to show some of the vast amount of tiny detail on the various parts of the disk.

The markings are generally referred to as light and dark "spots," although "festoons" and "ovals" are some of the other terms used from time to time. All these, as well as generally darker and lighter regions, can be followed as they are carried round the planet, sometimes speeding up or slowing down as they change their positions in the atmosphere. Some of the tiny markings may be seen for only a few days before they fade and disappear. The famous Great Red Spot has probably persisted for hundreds of years, but not without changes. Its position is usually detectable with most amateur-sized telescopes, but do not be too disappointed if you find it without the vivid coloration of some photographs and spacecraft images.

One of the advantages that Jupiter offers is that oppositions occur at intervals of about 13 months (roughly twice as often as those of Mars), and allow several weeks of observation (Table 10.1). Jupiter, too, is unlike Mars in that its apparent size does not vary greatly. Its phase and

▶ Fig. 10.1 The belts, zones, and other features that can be observed on Jupiter.

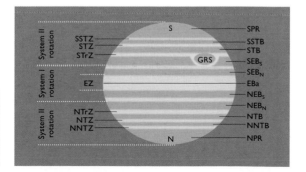

KEY		
B	belt	S	south
Ba	band	T	temperate
E	equatorial	Tr	tropical
N	north	Z	zone
PR	polar	GRS	Great
	region		Red Spot

TABLE 10.1: OPPOSITIONS OF JUPITER, 2012–2023			
Date	Declination	Diameter	Magnitude
2012 Dec 03	+21° 21′	49″	−2.4
2014 Jan 05	+22° 40′	47″	−2.2
2015 Feb 06	+16° 28′	46″	−2.1
2016 Mar 08	+5° 59′	44″	−2.0
2017 Apr 07	−5° 42′	44″	−2.0
2018 May 09	−16° 04′	45″	−2.0
2019 Jun 10	−22° 26′	46″	−2.1
2020 Jul 14	−21° 55′	48″	−2.3
2021 Aug 20	−13° 33′	49″	−2.9
2022 Sep 26	−0° 01′	50″	−2.9
2023 Nov 03	+13° 38′	50″	−2.9

tilt are negligible, so that when you are making sketches you do not have to worry about these details. Only if you are preparing outline blanks for whole-disk drawings must you take its flattening into account. The large size of Jupiter also makes it an ideal subject for planetary photography, and particularly striking results have been obtained by many amateurs using CCD equipment (Fig. 10.2).

Jupiter's rotation is fast – it is the cause of the considerable flattening – and the period of the central core amounts to about 9 hours 55 minutes 30 seconds. But the deep atmosphere and highly complicated meteorology mean that the atmospheric layers rotate at rather different speeds. The apparent "day" amounts to about 9 hours, 50 minutes 30 seconds in the equatorial region and 9 hours 55 minutes 40 seconds for the rest of the planet. (Slightly more accurate figures

◄ Fig. 10.2 A CCD image of Jupiter by Damian Peach, showing great detail, with the Great Red Spot and three White Ovals in the south, and one White Oval in the northern hemisphere.

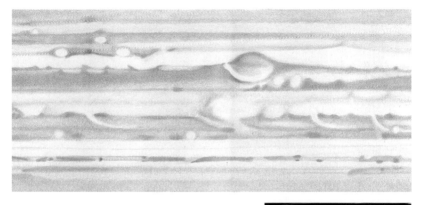

▲ Fig. 10.3 Part of a strip map of Jupiter, drawn by Richard McKim in 1988.

▶ Fig. 10.4 Whole-disk drawing of Jupiter, by Richard McKim.

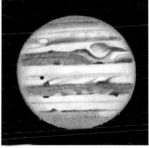

are used for calculation purposes.) The two visible divisions, and the two periods, are known as System I and System II, respectively. The yearly handbooks give tables showing the longitudes, in both of these Systems, for the center of the disk at any date and time.

There is so much detail visible on Jupiter when viewed through a moderate-sized telescope that it really is best to start by making sketches of individual features. The rotation periods of Jupiter are such that a different region is visible at the same time each night. The rotation is so fast that many observers have developed the technique of drawing "strip-sketches." These show just one or two particular belts and zones around the planet, the various markings being recorded as the rotation brings them into view (Fig. 10.3).

It is certainly worth attempting to make whole-disk drawings from time to time, because these then show the general aspect of the planet, which can change considerably from one apparition to the next (Fig. 10.4). When you start observing, do not attempt to include too much detail on whole-disk drawings, as you only have 10 minutes at the very most to complete the sketch before the rotation changes the appearance. Concentrate on showing the relative strength of the belts and zones. Just as with the other planets, intensity estimates are very useful in deciding the prominence of the various regions.

Jupiter, with its wealth of markings, is the best subject for central meridian transit timings. Simply record the times (to an accuracy of 1 minute) when features appear exactly in the center of the disk.

You can then work out the precise longitude of any feature quite easily by using the published tables. You only have to make sure that you can recognize the correct System to which a particular belt belongs. Sometimes this is a little difficult to decide when markings are on the borders of the two regions, but if in doubt, calculate and record both longitudes. If a feature is observed more than once, the longitudes can be compared, and a plot will show you how the markings change their positions over a period of time, overtaking one another as they are carried round the planet.

The most prominent feature, the Great Red Spot, is no exception to the general drift in longitude, and has been followed many times completely "round" the planet. Its size, prominence, and color are all subject to change, and it has been gradually fading and shrinking. A second, smaller red spot appeared unexpectedly in 2006 March.

Jupiter's satellites

Jupiter's four major satellites, Io, Europa, Ganymede, and Callisto, have been known since Galileo first turned his primitive telescope on to the planet. (To this day they are frequently called the "Galilean satellites.") They are visible with even the smallest binoculars, and it is said that they may be visible to the naked eye, especially to an observer in the tropics where the planet may be overhead. They weave a complex pattern around Jupiter, and frequently one or more may be hidden behind the planet. It is only on rare occasions, however, that all four appear to be "missing." Most yearbooks give diagrams showing the relative positions at any one time, from which the individual satellites may be identified.

Apertures of 50–75 mm are enough to show various interesting events as the satellites pass in front of and behind the planet (Table 10.2). Their shadows on Jupiter's clouds can be seen, as well as the transits of the bodies themselves, although the latter are much harder to detect. The satellites are eclipsed when they pass into Jupiter's shadow, and may also be occulted by the body of the planet. What is more, every six years, when their orbital plane is aligned with the Earth, they also affect one another in the same way. These satellite phenomena are fascinating to watch, and should be timed as accurately as possible. Although no detail can be seen on any of the satellites with even very large telescopes, the difference in their appearance is quite striking when they are seen against the

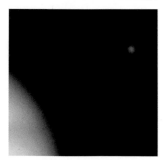

▲ *Fig. 10.5 A remarkable image of Jupiter and Ganymede, obtained by Steve Massey.*

TABLE 10.2: JUPITER'S SATELLITE PHENOMENA	
Transit	Satellite passes in front of Jupiter
Shadow transit	Satellite's shadow cast on Jupiter's clouds
Eclipse	Satellite disappears in Jupiter's shadow
Occultation	Satellite passes behind Jupiter

disk of the planet. One of the achievements of visual astronomy was the detection (in 1922) of Ganymede's captured rotation by a great observer, T. E. R. Philips. Nowadays, a few amateur astronomers are even obtaining CCD images that reveal detail on that satellite.

▲ *Fig. 10.6 Image by David Hanon of Jupiter and Io, together with Io's shadow transit of the disk.*

Saturn

With its spectacular system of rings, Saturn is as striking an object to observe as Jupiter (Fig. 10.7). The methods used are essentially the same, although Saturn only rarely shows distinct markings, apart from the belts and zones (Fig. 10.8). This is partly because of its greater distance, but we now know that a high atmospheric haze hides the features in the lower cloud layers. This makes it all the more important to follow the various light and dark spots that do sometimes arise. Large, persistent white spots seem to occur occasionally when Saturn is close to perihelion.

Saturn's oppositions are separated by intervals of about 378 days (Table 10.3). Although the phase is small, amounting to no more than 6° at the most, the most obvious variation comes in the tilt of the rings

▶ *Fig. 10.7 A beautiful CCD image of Saturn by Damian Peach, clearly showing the banded structure of the atmosphere and fine detail in the rings.*

TABLE 10.3: OPPOSITIONS OF SATURN, 2012–2023			
Date	Declination	Diameter	Magnitude
2012 Apr 15	−07° 31′	19″	+0.4
2013 Apr 28	−11° 42′	19″	+0.3
2014 May 10	−15° 21′	19″	+0.3
2015 May 23	−18° 21′	19″	+0.2
2016 Jun 03	−20° 35′	19″	+0.2
2017 Jun 15	−21° 58′	19″	+0.2
2018 Jun 27	−22° 27′	19″	+0.2
2019 Jul 09	−22° 01′	18″	+0.3
2020 Jul 20	−20° 39′	19″	+0.3
2021 Aug 02	−18° 26′	19″	+0.2
2022 Aug 14	−15° 26′	19″	+0.3
2023 Aug 27	−11° 46′	19″	+0.4

(and of the planet, of course). Twice in a Saturnian year (once every 15 years), the Earth passes through the plane of the rings. The rings are extremely thin, with an average thickness of possibly less than 100 meters (330 feet), and when this occurs they may temporarily disappear. Gradually, more of one hemisphere of the planet comes into view, whilst the rings begin to hide the other, until the tilt reaches its maximum value of about 28°. Then the change reverses direction, until the Earth is again in the ring-plane, after which the other hemisphere is fully exposed. The last passages of the ring-plane occurred in 1995 and 2010, the northern hemisphere being visible now.

This constantly changing aspect, together with the planet's very appreciable polar flattening, means that the preparation of blanks for drawings is very complicated, the elliptical shape of both the planet and the rings having to be correctly rendered. You will probably find it best to obtain master blanks from one of the amateur observational organizations, unless you are very good at drafting such material.

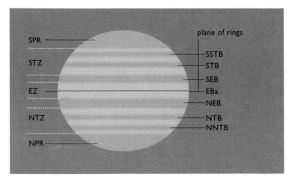

◄ Fig. 10.8 The belts, zones, and polar regions of Saturn.

KEY			
B	belt	S	south
Ba	band	T	temperate
E	equatorial	Tr	tropical
N	north	Z	zone
PR	polar region		

The rings of Saturn

The main portions of the rings – the outer, reasonably bright Ring A, the even brighter Ring B, and the Cassini Division between them – are easy to see. Ring C, the innermost, is not so readily visible, and its transparent nature gave rise to its alternative name of the Crêpe Ring. When viewed against the disk it can look similar to a dusky belt. Frequently the shadow of the planet obscures a portion of the rings, and they may

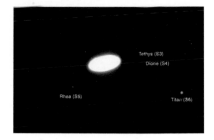

▲ *Fig. 10.9 An image of Saturn (greatly overexposed) and the four brightest satellites.*

appear dark against the disk when, near the dates of ring-plane passage, the Sun and Earth are on opposite sides of their plane. Various irregularities are reported from time to time, and should be carefully drawn.

Satellites

Again like Jupiter, Saturn has an interesting set of satellites, the brightest, Titan, being of magnitude 8 (Fig. 10.9). Three more (Rhea, Tethys, and Dione) are brighter than 10.5, and another (Enceladus) over 12.0. Iapetus varies between about magnitude 10.2 and 11.9, because one hemisphere is much darker than the other. It is always brightest at its western elongations. When the Earth passes through the satellites' orbital plane they may appear as bright beads of light strung on the thin line of the rings. In large telescopes, satellite phenomena like those of the Galilean satellites of Jupiter are then visible. Diagrams showing the positions of the brightest satellites are published in some yearbooks.

The outer planets

Unfortunately the outer planets do not give very much scope for observation, but they do offer the challenge of locating and following them. Uranus and Neptune are reasonably bright at opposition (reaching maximum magnitudes of around 5.6 and 7.7 respectively, although normally less than this). They are not too difficult to find in small telescopes, using suitable charts and ephemerides. They slowly work their way eastward against the stars (Fig. 10.10), Uranus increasing its RA by about 20 minutes per year, and Neptune by about half that amount. Both are at present in the southern part of the ecliptic (lying in Pisces and Aquarius, respectively). Because of their distance from the Sun (and Earth) their paths against the sky have relatively large retrograde loops.

▲ Fig. 10.10 The motion of Uranus, Neptune, and Pluto in the sky over a period of 10 years, omitting their retrograde loops.

◀ Fig. 10.11 A CCD image by Ed Grafton of featureless Uranus (seen almost pole-on) and the four brightest satellites.

Uranus

Uranus is distinctly different from stars in appearance, and shows a minute disk in a good telescope, which may be seen as bluish or greenish depending on the observer's eyesight. Faint belts and zones have been recorded by people using very large telescopes, but there are no obvious, definite markings, and this was confirmed by the Voyager 2 probe during its fly-by. The planet is very unusual because its axis of rotation is nearly in the orbital plane, being tilted by about 98°, so that at times (in 1986 for example) the polar regions are presented to the Earth. When this happens little detail can be expected. Table 10.4 gives forthcoming oppositions of Uranus, together with its declination, diameter, and magnitude. As can be seen, both the diameter and magnitude change extremely slowly.

Uranus has at least 27 satellites, but most of these are extremely small and faint. All were once beyond the reach of amateurs, but in recent years, CCD equipment has enabled some of the larger and brighter satellites to be captured by amateurs (Fig. 10.12).

Neptune

Neptune shows no detail in amateur telescopes (Fig. 10.12) but is moderately bright, being visible in binoculars, so its dates of opposition are given in Table 10.5. Both Uranus and Neptune have been suspected of showing some changes in magnitude over a very long period of time. There have been suggestions that variations might be related to solar activity, but this remains quite uncertain. The magnitudes of both planets

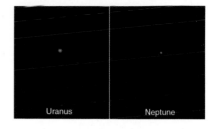

▲ *Fig. 10.12 A comparison of images of Uranus and Neptune, obtained with the same telescope and equipment.*

could be estimated quite easily by variable star methods. The values given in the tables assume that the planets' albedo remains constant.

Pluto

Pluto, a body now classed as a dwarf planet, has an eccentric orbit that occasionally brings it inside the orbit of Neptune – and is very difficult to locate with most amateur telescopes. The magnitude is about 14, so apertures of about 300 mm are required for it to be visible at all. Being so faint, it is easy to confuse Pluto with the surrounding stars, and the best method of locating it is by means of photography on different dates. Charts are published in some of the astronomical yearbooks. At present the yearly motion is similar to that of Neptune, because both bodies are at approximately the same distance from the Sun. (Pluto's distance began to exceed that of Neptune in 1999.)

TABLE 10.4: OPPOSITIONS OF URANUS, 2012–2023			
Date	Declination	Diameter	Magnitude
2012 Sep 29	+1° 55'	3.60"	+6.06
2013 Oct 03	+3° 30'	3.60"	+6.06
2014 Oct 07	+5° 04'	3.61"	+6.05
2015 Oct 12	+6° 38'	3.61"	+6.05
2016 Oct 15	+8° 10'	3.62"	+6.04
2017 Oct 19	+9° 40'	3.62"	+6.03
2018 Oct 24	+11° 08'	3.63"	+6.02
2019 Oct 28	+12° 34'	3.64"	+6.01
2020 Oct 31	+13° 57'	3.65"	+6.00
2021 Nov 04	+15° 16'	3.76"	+5.60
2022 Nov 09	+16° 32'	3.77"	+5.60
2023 Nov 13	+17° 43'	3.78"	+5.60

Date	Declination	Diameter	Magnitude
TABLE 10.5: OPPOSITIONS OF NEPTUNE, 2012–2023			
2012 Aug 24	−11° 26′	2.52″	+7.65
2013 Aug 27	−10° 42′	2.52″	+7.64
2014 Aug 29	−9° 56′	2.52″	+7.64
2015 Sep 01	−9° 10′	2.53″	+7.64
2016 Sep 02	−8° 24′	2.53″	+7.64
2017 Sep 05	−7° 36′	2.53″	+7.64
2018 Sep 07	−6° 48′	2.53″	+7.64
2019 Sep 10	−5° 59′	2.53″	+7.64
2020 Sep 11	−5° 10′	2.53″	+7.64
2021 Sep 14	−4° 20′	2.36″	+7.80
2022 Sep 16	−3° 30′	2.36″	+7.80
2023 Sep 19	−2° 39′	2.36″	+7.80

Pluto is currently in Sagittarius, farther away from the ecliptic than Uranus and Neptune, because of its high orbital inclination. Brightness changes occur as the object approaches and recedes from the Sun, and are thought to be linked with changes in the relative amounts of methane frost on the surface and methane gas in the tenuous atmosphere. The discovery of its largest satellite, Charon, revealed that Pluto's rotation period and Charon's orbital period are retrograde, and identical at 6 days 9 hours 22 minutes. Mutual eclipses have helped to establish the presence of large-scale features on Pluto. It is now known to be one of a large population of icy bodies (possibly amounting to many thousands, including the larger body called Eris) orbiting at the edge of the Solar System, and forming what is known as the Kuiper-Edgeworth Belt.

Comets

Comets are often referred to as "dirty snowballs." As far as we know, they are very small bodies only a mile or so across, consisting of ices and dust particles. They remain invisible until their elongated orbits bring them into the center of the Solar System, where the heat of the Sun causes some of the ices to evaporate as various gases. Some comets may then become spectacular objects, such as Comet Hyakutake in 1996 (Fig. 10.13) and Comet Hale-Bopp in 1997 (Fig. 10.16).

Never miss the opportunity to observe a bright comet. Although several are usually visible in a year to amateurs with moderate-sized equipment, most of those that are regular visitors to the inner Solar System are very faint. The only exception is Halley's Comet, which normally becomes quite prominent each time it returns in its 76-year orbit. The really spectacular comets are the unexpected ones, which

▲ Fig. 10.13 A photograph of Comet Hyakutake by Akira Fujii, with rayed structure clearly visible within the blue gas tail.

may suddenly arrive from almost any direction, and which may be approaching the Sun for the very first time. There are then a few hectic weeks for cometary astronomers, while the bright comet swings round the Sun, before disappearing again into the distance, perhaps to return only in thousands of years' time. Occasionally, comets may fragment. In 2005–2006 Comet Schwassmann-Wachmann broke into dozens of pieces, each appearing as a small comet in its own right. Another instance was Comet Shoemaker-Levy 9, which broke up into 20-odd fragments and crashed into Jupiter in July 1994, producing dark scars at the impact sites that lasted for weeks (Fig. 10.14).

Comets frequently differ greatly in appearance. All show a head (the coma) which may never appear as more than a fuzzy patch, even when the comet is closest to the Sun. Others may show distinct features, the most notable of which is the tail. This usually becomes more conspicuous as the comet nears perihelion, and the increasing heat of the Sun releases both gases and dust from the icy cometary

▲ Fig. 10.14 A sequence of photographs by Thierry Legault, showing some of the scars left by the impact of the fragments of Comet Shoemaker-Levy 9 on Jupiter.

body. In general, tails point away from the Sun. Some comets show two tails: one frequently curved or fan-shaped, and formed of dust particles lying in the plane of the comet's orbit; and the other more-or-less straight, gaseous, and always pointing in the opposite direction to the Sun. Although not necessarily apparent to the eye, the two types of tail have different colors. Dust tails have a yellowish tinge (because their color comes from reflected sunlight), whereas gas tails are always blue. During part of its passage, Comet Hale-Bopp showed these two types of tail particularly clearly (Fig. 10.16).

The appearance of cometary tails may vary greatly depending upon the direction in which they are being viewed. This was the case with Comet Hale-Bopp, which at one stage showed a radiating structure (Fig. 10.15). A comet may occasionally show an antitail – a "spike" of dust that seems to point toward the Sun. This is actually an optical illusion: because of the relative positions of the Earth and comet, we are looking through part of the tail, such that the particles appear projected against the sky "ahead" of the comet. A few cases of multiple (dust) tails are known, and tails have been recorded as stretching half-way across the sky. The dust from comets is thought to give rise to most of the tiny particles that are observed as meteors and which form the zodiacal light. The straight gas tails often consist of several rays, which may sometimes show distinct kinks or breaks. These are known as discontinuity events and are an indication of a change in the polarity of the interplanetary magnetic field at that point.

Within the coma, a tiny, bright, star-like point (the nucleus) is some-times visible, and this looks the same even with the highest magnifica-tion. (Cometary nuclei are too small to appear as anything other than points in even the largest telescopes.) This nucleus is the dense

▲ Fig. 10.15 Radiating tail structure as seen in Comet Hale-Bopp, photographed by A. Kelly.

▶ Fig. 10.16 The spectacular gas and dust tails of Comet Hale-Bopp, photographed by Akira Fujii.

► *Fig. 10.17 Detailed structure within the coma of Comet Hale-Bopp, as drawn by Richard McKim on 1997 April 17.*

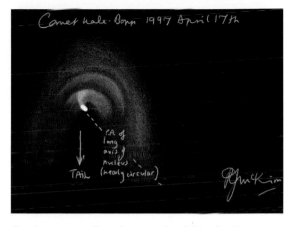

cloud of material immediately surrounding the actual solid body. Do not confuse it with the structure that you may sometimes see in the coma itself, and which appears as shells or jets of material.

Observing comets

You can use any equipment for observation, but it is often easier to detect the extent of tail with the naked eye, binoculars or wide-field telescopes rather than with larger telescopes. Large apertures and magnifications are required, however, to see the intricate detail near the nucleus. With any equipment, make drawings in ways similar to those used for rendering the planets (Fig. 10.17). It is also worth trying some photography. This will record both the actual position of the head, and may capture the tail or some of its structure when it is too faint to be seen by other means. Wide-field, fast lenses are required for this work, while telescopes or long-focus cameras are needed to record details of the coma. To obtain the best results the equipment should be guided to follow the motion of the comet itself, when the background stars will show as trails. Short-focus lenses can be driven to follow the stars.

Many advanced amateurs undertake comet searches, but this requires an immense degree of patience to learn the star patterns and possible confusing objects such as clusters and galaxies over a large area of the sky. Wide-field, large binoculars or short-focus telescopes are normally employed for this sort of work, and searches are often carried out in the region near the Sun where a comet may approach very close to the Sun and Earth without being detected. Several comets have been discovered by amateurs using the Internet to search the archive of photographs of the Sun that have been obtained by the SOHO satellite, in a method reminiscent of the discovery of comets at total solar eclipses.

THE STARS

The colors of stars are an approximate indication of the temperatures of their visible surfaces. These temperatures range from about 40,000°C for rare blue-white stars like β Orionis (Rigel) down to about 3000°C for a deep red star such as μ Cephei, the famous "Garnet Star." There are even rare examples beyond these two extremes. Some interesting colored stars are given in Table 11.1, but some observers may have difficulty in seeing the tints, because much depends on the equipment and the observer's eyesight. At low light levels colors are not readily visible to the eye, although they may be distinct on photographs. Greater apertures make them more apparent. In addition, various observers have completely different responses, some being blue-sensitive and having difficulty with red stars, and others finding the opposite. Generally red stars seem to become brighter and brighter the longer they are observed. (This is a problem with some variable-star observations, where special precautions must be taken against the effect.) Double stars often show striking color combinations, mainly caused by the effect of contrast.

Spectral classes

Catalogs frequently list the spectral classes of individual stars. This is a more scientific method of describing temperature and composition, defined on the basis of which elements cause the lines that are visible in the spectrum of a particular star. The classes, arranged from hottest to coolest, follow the now rather jumbled sequence O, B, A, F, G, K, M. (The standard mnemonic is "Oh Be A Fine Girl Kiss Me.") Recently, an additional class, L, has been added at the cool end

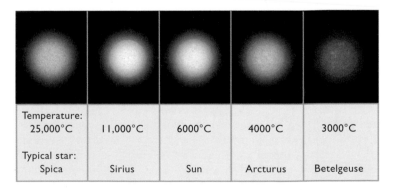

Temperature:				
25,000°C	11,000°C	6000°C	4000°C	3000°C
Typical star:				
Spica	Sirius	Sun	Arcturus	Betelgeuse

▲ Fig. 11.1 The temperatures and colors of certain bright, well-known stars. The diameters are not to scale: Betelgeuse is larger than the orbit of Mars.

TABLE 11.1: COLORED STARS		
Star	Name	Color
α Aur	Capella	Yellow
α Boo	Arcturus	Orange-yellow
α CMa	Sirius	Blue-white
μ Cep	"Garnet Star"	Deep red
α Lyr	Vega	Blue-white
β Ori	Rigel	Blue-white
α Sco	Antares	Red
α Tau	Aldebaran	Orange

▼ *Fig. 11.2 R Leporis is one of the reddest stars in the sky.*

to allow for a group of stars with extremely low temperatures. The rarer classes R, N, S, and C (which are cool stars), and WN and WC (hot stars) are sometimes mentioned. Most of the classes have 10 main subdivisions, numbered 0–9 (hottest to coolest), and the vast majority of stars range from about O5 to M8. The Sun is a G2 star, surface temperature about 6000°C.

Stars are also described on the basis of their sizes and luminous output. For historical reasons, the majority of stars (like the Sun) are known as dwarfs, larger stars being subgiants, giants or supergiants. Overall, stars range from supergiants like α Scorpii (Antares), larger than the orbit of Mars, and μ Cephei, which is actually larger than the orbit of Saturn, to tiny white dwarfs smaller than the Earth. A few interesting examples are noted in Table 11.2. (There are other types of star, of course, most notably brown dwarfs, which are extremely cool and faint, and neutron stars, just 10–20 km (6–12 miles) across, but these are beyond amateur observation or study.) Although supergiant stars are often cool, they have such a large surface area that they are frequently extremely luminous, and have high absolute magnitudes.

Magnitudes and distances of stars

To the eye, a star of first magnitude appears twice as bright as one of second magnitude, which in turn appears twice as bright as one of third magnitude. Actually they are not – each magnitude interval is slightly brighter than that amount, the true ratio being 2.512:1. The eye and brain perceive that constant *ratio* as a constant *step*. For most purposes we can forget the mathematical relationship, although it helps to remember that a first-magnitude star is exactly 100 times the brightness of a star of sixth magnitude. It is sometimes useful to know the magnitudes of some of the brightest stars, and a number of these are given in Table 11.2. The very brightest have negative magnitudes. Some exact, fainter magnitudes are given in the small charts for Ursa Minor, Crux, and the Pleiades (Fig. 11.3), on page 173.

TABLE 11.2: APPARENT AND ABSOLUTE MAGNITUDES, AND SPECTRAL CLASSES			
	m	M	Spectral class
α Aql	+0.76	+2.20	A7 subgiant/dwarf
α Aur	+0.08	−0.48	G2 giant and G6 giant
α Boo	−0.05	−0.31	K2 giant
α CMa	−1.45	+1.41	A0 dwarf
α CMi	+0.40	+2.68	F5 subgiant
α Car	−0.62	−5.53	F0 supergiant
α Cen	−0.01	+4.34	G2 dwarf
β Cen	+0.61	−5.43	B1 giant (variable)
α Cru	+0.77	−4.19	B0.5 subgiant and B1 dwarf
α Cyg	+1.25	−8.23	A2 supergiant
α Eri	+0.45	−2.77	B3 dwarf
α Lyr	+0.03	+0.58	A0 dwarf (variable)
α Ori	+0.45	−5.14	M2 supergiant
β Ori	+0.18	−6.69	B8 supergiant
α Sco	+1.06	−5.28	M1 supergiant and B2.5 dwarf
α Tau	+0.87	−0.63	K3 giant
α Vir	+0.98	−3.55	B1 dwarf
Sun	−26.8	+4.79	G2 dwarf

The distances of stars have always been very difficult to measure. The results obtained by the Hipparcos satellite were a major breakthrough in this respect and have enabled us to gain a far better understanding of the distribution of stars in our region of the Galaxy. Stellar distances are so great that kilometers or miles and even astronomical units are inconvenient, so astronomers use either light-years or, for preference, parsecs (pc). One parsec (parallax second) is that distance at which the radius of the Earth's orbit, 1 astronomical unit, subtends an angle of 1 second of arc. It is 3.216 light-years, 206,265 astronomical units or about 31 million million km. Kiloparsecs (kpc) and Megaparsecs (Mpc) – 1000 and 1 million parsecs, respectively – are also used for galactic and extragalactic distances.

The magnitudes mentioned already – those that we see from Earth – are apparent magnitudes (m). But stars differ greatly in their actual brightness and an apparently bright star may be a brilliant one far away, or a faint one near at hand. So the true brightness of stars (the luminosity) is calculated as if they were all located at a standard distance. This has been chosen as 10 pc, and the magnitudes are known as absolute magnitudes (M). The difference between the two types of magnitude may be very striking and a few notable examples (such as α Cygni) can be seen in the table.

▶ Fig. 11.3 Visual stellar
magnitudes in three areas
of the sky.

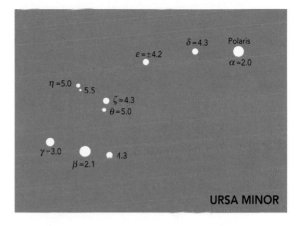

▶ Ursa Minor. Note
that Polaris itself is a
Cepheid-type variable,
but its variations are
too small to be followed
visually.

▶ The Pleiades, M45.
This very rich cluster
contains hundreds
of stars, but only the
brightest are shown here.

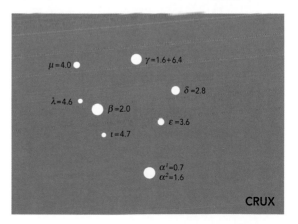

▶ Crux is visible for
most southern observers,
but is poor in readily
recognizable faint stars.

Variable stars

Many stars are variables and show changes in their brightness. A plot of a star's apparent magnitude against time produces a light-curve, and this can give a lot of information about the star itself. Depending upon the type of object, the changes may take place in just a few minutes or over a period of many years. The most obvious cause of variation is when stars are part of a binary system, where the orbital plane is aligned with the Earth, so that the two stars occasionally eclipse one another. Algol (β Persei) is the most famous example of this class, ranging between about magnitudes 2.2 and 3.4. In some cases both primary and secondary minima are observed (Fig. 11.4, top), as the bright and faint components, respectively, are eclipsed.

Apart from these eclipsing systems there are many other types of variables, some of which are close binary systems and others single stars. Many forms of variation are linked with particular stages of stellar evolution. Some stars (particularly those that pulsate) exhibit extremely regular variations, the best known being the Cepheids, where there is a relationship between the period and the absolute luminosity, which means that these stars are a vital link in determining

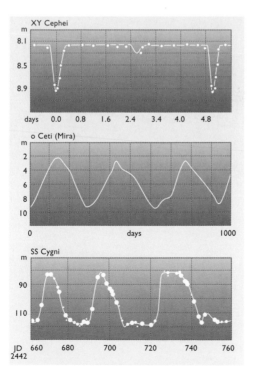

◀ Fig. 11.4 Top: The eclipsing binary light-curve is drawn from single observations (dots). Center: The averages from many individual observations are used to construct the smooth light-curve of a long-period variable. Bottom: The size of the dots gives an indication of the number of observations used at each point in this light-curve of an eruptive variable.

stellar distances. (The changes of the brightest Cepheid, l Carinae, are clearly visible to the naked eye.)

For amateur astronomers the most important types are probably the long-period variables (LPVs), semiregulars (SR), and various eruptives (which include some that show sudden fades rather than outbursts). There are so many variable stars, of all magnitudes, that their study is one of the most satisfying fields of research for amateurs, who perform a valuable scientific function by keeping stars under long-term surveillance – in some cases extending over more than 100 years. They are able to monitor objects and alert professional astronomers of any significant or sudden changes. The professionals may then be able to justify devoting precious, large-telescope time to studying the objects concerned. A few interesting objects are given in Table 11.3, where "P" indicates "Period."

TABLE 11.3: SOME INTERESTING VARIABLE STARS					
Designation	RA	Dec. (2000)	Range (mag.)	Type	Remarks
R And	00h 24m	+38° 35'	6.9–14.3	LPV	P = 409 days
ε Aur	05h 02m	+43° 48'	3.5–4.5	Eclipsing	P = 27 years
ζ Aur	05h 02m	+41° 04'	5.0–5.5	Eclipsing	P = 972 days
l Car	09h 45m	−62° 31'	3.3–4.2	Cepheid	P = 35.54 days
R Car	09h 32m	−62° 47'	3.9–10.0	LPV	P = 309 days
γ Cas	00h 58m	+60° 43'	1.6–3.0	Irregular	–
ρ Cas	23h 54m	+57° 30'	4.1–6.2	SR (pec)	–
R Cen	14h 17m	−59° 55'	5.4–11.8	LPV	Deep and shallow minima, P = 546 days
T Cep	21h 10m	+68° 29'	5.4–11.0	LPV	P = 388 days
δ Cep	22h 29m	+58° 26'	3.9–5.0	Cepheid	P = 5.67 days
μ Cep	21h 44m	+58° 47'	3.6–5.1	SR	P = 730 days
ο Cet	02h 19m	−02° 58'	3.5–9.1	LPV	"Mira," P = 332 days
R CrB	15h 49m	+28° 10'	5.8–14.8	RCB	Irregular fades
SS Cyg	21h 43m	+43° 35'	8.2–12.4	UG	Eruptive star
ψ Cyg	19h 51m	+32° 55'	5.2–13.4	LPV	P = 408 days
AC Her	18h 30m	+21° 52'	7.0–8.4	RV	Deep and shallow minima
R Hya	13h 30m	−23° 17'	4.5–9.5	LPV	P = 389 days
R Leo	09h 48m	+11° 26'	4.4–11.3	LPV	P = 310 days
β Lyr	18h 50m	+33° 21'	3.3–4.3	Eclipsing	P = 12.91 days
U Mon	07h 31m	−09° 47'	5.9–7.8	SR	–
W Ori	05h 05m	+01° 11'	6.0–7.7	SR	–
k Pav	17h 57m	−67° 14'	3.9–4.9	Cepheid	P = 9.09 days
β Per	03h 08m	+40° 58'	2.1–3.4	Eclipsing	"Algol," P = 2.87 days
L2 Pup	07h 14m	−44° 38'	2.6–6.2	SR	–

▲ Fig. 11.5 At maximum (left) the long-period variable Mira (o Ceti) is approximately magnitude 3. At minimum (right), it is almost magnitude 10.

The long-period variable Mira (o Ceti) is typical of its class, with an average period of about 332 days, which is, however, subject to fluctuations. It is shown bright and faint in Figure 11.5, and its light-curve in Figure 11.4, center. Related stars are the semiregulars, of which there are several subclasses, exhibiting varying degrees of regularity. Also similar are the RV-Tauri stars, which show alternating primary and secondary (that is, deep and shallow) minima, which occasionally interchange places. A chart and comparisons for AC Herculis, a good example of this type, are shown in Figure 11.6. Another particularly interesting (and rare) class is that of the R Coronae Borealis stars,

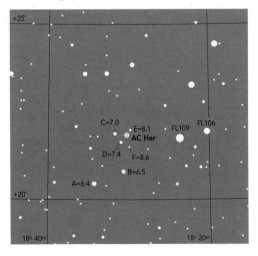

◄ Fig. 11.6 A chart (with comparison star designations and magnitudes) for the semiregular variable AC Herculis.

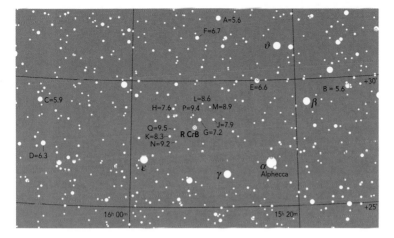

▲ Fig. 11.7 A chart (with comparison star designations and magnitudes) for R Coronae Borealis, which exhibits unpredictable fades.

which show sudden, unpredictable fades, sometimes by several magnitudes, and which may take months or even years to recover to their normal brightness. A chart for the type star (R CrB) is shown in Figure 11.7. This star should be checked on every possible occasion (using binoculars) and any fade reported immediately.

Estimating magnitudes

It is not difficult to estimate the magnitudes of stars. You need to have the magnitudes of comparison stars and can obtain these on special charts issued by variable-star organizations, frequently with additional "finder" charts to help in locating the variable (similar to Figs. 11.6 and 11.7). Once you have found the field, see if the variable is visible. If not, make a note of the faintest comparison star that you can see, and write down "not visible, below ....." Even that information is helpful. If the star is visible, decide which of the comparison stars seems slightly brighter and which seems fainter. If the variable appears exactly the same as one of them, check the next brighter or fainter as well. It is usually quite easy to get this far, and you have already roughly determined the star's magnitude.

Try using the "fractional method" to take it further. Look at the three stars again. Is the variable halfway between the others in brightness? If so, write down "A(1)V(1)B," where A and B are the bright and faint comparisons, respectively. "V" is entered for the variable (whatever its actual name may be), and the figures represent the fractions – exactly the same in this case. The sequence is always "bright star, fraction, variable, fraction, faint star." Does the variable seem one-third of the way from one to the other? Then write "C(1)V(2)D," representing "C, one-third, variable, two-thirds, D." Other fractions could be given by "...(1)V(3)...," "...(3)V(2)...," "...(4)V(1)...," or whatever the case may warrant. Do not try to divide the interval into more than five parts, because errors then begin to creep in. Careful scientific tests have shown that – despite what some observers have claimed – the human eye and brain cannot differentiate between ratios of (say) 4:1 and 5:1. Dividing the interval between two comparison stars into 10 parts, as has sometimes been suggested in the past, is attempting the impossible.

Other methods of making estimates can be used when you have gained experience. The most notable of these is the "Pogson method" in which the eye is trained to recognize differences in terms of one-tenth-magnitude steps. This method is particularly useful when only one comparison star is available.

You can obtain the actual, or deduced magnitude by simple arithmetic. Take the difference between the two comparison stars, and work out one (or both) fractions. Remember that magnitude values increase for fainter stars, so add the fraction to the magnitude of the bright comparison, or subtract from that of the fainter, whichever is the easiest. Give the result to 0.1 magnitude.

This all sounds more complicated than it is in practice. Try it, and you will be pleasantly surprised. There are some problems, of course. Do not stare at red stars, otherwise they will seem to get brighter and brighter. Take short glimpses instead. (Estimates of red stars always show far greater difference between observers than with bluer objects.) Try turning your head, because of two equal stars, the one that is closer to the nose or lower in the field of view (or both) will always appear slightly brighter.

In making estimates of stars that are close to the limit of visibility, variable-star observers frequently use averted vision, looking slightly to one side of the star in question. (When looking directly at a star, the image actually falls on a slightly less sensitive portion of the retina.) This technique requires considerable practice, because it is important that the images of both the variable and the comparison star (or stars) are brought in turn to the same region of the retina.

Dwarf novae and novae

There are many forms of eruptive variables (mostly close binary systems) and outbursts of a large number of individual objects may be seen quite frequently, even though at irregular intervals. One type of eruptive, known as a dwarf nova, is very popular with amateurs. These stars show outbursts of a few magnitudes – and thus show distinct

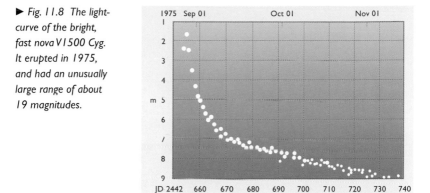

▶ Fig. 11.8 The light-curve of the bright, fast nova V1500 Cyg. It erupted in 1975, and had an unusually large range of about 19 magnitudes.

changes – at irregular intervals. The eruptions are unpredictable but, when averaged over time, each star exhibits its own characteristic mean interval. One favorite dwarf nova is SS Cygni, whose light-curve is shown in Figure 11.4, bottom.

Novae are an extreme form of eruptive star, sometimes rising by 10 magnitudes (10,000 times) or even more in just a couple of days. However, it is not until a particular star erupts that it is even known to exist, so the outbursts are quite unpredictable. Once the star has been discovered it can be followed by the ordinary methods of estimating magnitudes, although there is usually an initial problem in finding suitable comparison stars and obtaining their magnitudes. A particularly notable nova was Nova Cygni 1975 (Fig. 11.8), which had the exceptional rise of about 19 magnitudes, and was so bright that it was "discovered" by dozens of people all over the world (Fig. 11.9).

It is obviously most important to discover these objects as soon as possible, so many amateurs undertake nova patrols, either visually or photographically. Like comet searching, it requires considerable patience to learn the stellar patterns over even a small area of the sky. All the other variable stars must also be recognized to prevent numerous "false alarms." Photographs have the advantage of providing a permanent record, but should be taken in pairs to enable the inevitable emulsion faults to be detected. They also have to be developed and examined immediately if a nova is to be detected at an early stage.

Novae most frequently occur in the regions closest to the plane of the Milky Way, and various coordinated patrols keep close watch over these areas. Both these patrols and individual observers have recorded some particular successes over the years, most notably the UK Nova and Supernova Patrol, coordinated by *The Astronomer* magazine. The information that they have obtained has frequently

◀ Fig. 11.9 Nova V1500 Cyg at magnitude 2 (left) and at about magnitude 11 (right), many weeks into its decline.

been of vital interest to professional astronomers. It is most important that observers submit details of any suspected object to a patrol coordinator, who will then obtain confirmation before forwarding details to the central clearing-house for astronomical discoveries, the Central Bureau for Astronomical Telegrams (CBAT) in Cambridge, Massachusetts. This ensures that false alarms do not cause professional astronomers to devote precious observing time to non-existent "discoveries."

Double and multiple stars

Many stars appear close to one another in the sky. Some of these have no actual connection, being at greatly different distances and merely lying on the same line of sight. These are optical doubles. In other cases the stars are in orbit around one another, and form true binary systems. Multiple systems consisting of three or more stars also occur. Many doubles (of both types) are striking objects in binoculars or small telescopes. One or two, such as ζ Ursa Majoris (Mizar) with its companion Alcor, are reasonably easy for the naked eye. Eyesight alone will just about resolve ε Lyr, but this is easily split with binoculars. In a small telescope with magnifications of 100–200, it is visible as four components, which is why it is often known as the "Double Double."

Although Mizar and Alcor are sufficiently wide to have been given separate names, most doubles or multiples appear so only with optical assistance. In these cases, the stars are normally lettered in decreasing order of brightness, giving (for example) "α Centauri A," "α Centauri B," and so on. The fainter star of two is frequently called the "comes" (the Latin for "companion," the latter term being often used instead). Lists of doubles and multiples usually give two values for each pair of stars, specifying the position of the faint star relative to the bright one. These values are the Position Angle (PA) of the line joining the two stars, and their separation, which is usually measured in seconds of arc (Fig. 11.10). Determination of these two parameters requires a specialized piece of equipment, a micrometer, explained later. It is probably best to start observing doubles and multiples (Fig. 11.11) purely for the sake of general interest, using published lists, and then to progress to making or purchasing a micrometer when one begins to develop a fascination with the subject.

Double stars are an excellent test of telescopic resolution, and some suitable objects are given in Table 11.4. However, do not be unduly disappointed if you seem to fail to reach the theoretical resolution of your telescope – once again, considerable practice and excellent seeing conditions are required. Note that in the table, for naked-eye doubles the position given corresponds to the first cataloged star. Again, caution should be exercised in accepting the magnitudes listed, although these are taken from the latest available catalog.

▶ Fig. 11.10 Position Angle (PA) is always measured in degrees from north through east. The separation is normally given in seconds of arc.

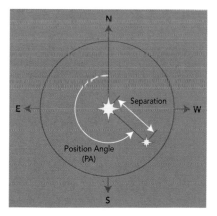

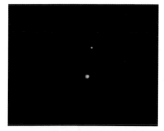

◀ Fig. 11.11 Contrasting colors in the double star γ Leporis, photographed by Ed Grafton.

TABLE 11.4: DOUBLE AND MULTIPLE STARS				
Designation	RA (2000)	Dec.	Separation (arcseconds)	Magnitudes; notes
γ And	02^h 04^m	+42° 21'	9.8 (A–B)	Multiple: 2.3, 5.1, 5.5, 6.3
ζ Aqr	22^h 29^m	−00° 02'	2.1	4.4, 4.6: white stars
γ Ari	01^h 54^m	+19° 18'	7.7	3.8, 3.9: white stars
ε Boo	14^h 45^m	+27° 04'	2.6	"Pulcherima": 2.5, 4.6; orange and blue
ζ Cnc	08^h 12^m	+17° 39'	6.0	5.0, 6.2; fainter third component
α CVn	12^h 56^m	+38° 19'	19.0	2.9, 5.6
η Cas	00^h 49^m	+57° 49'	12.7	3.5, 7.2; yellow and red stars
ι Cas	02^h 30^m	+67° 24'	7.1 (A–C)	4.6, 8.4; B = 6.9, A–B = 2.7
α Cen	14^h 40^m	−60° 51'	16.4	0.0, 1.4; yellow stars
66 Cet	02^h 13^m	−02° 24'	16.6	5.7, 7.6
δ^1, δ^2 Cha	10^h 45^m	−80° 29'	wide optical	5.6, 4.4; orange and blue
α Cru	12^h 27^m	−63° 06'	4.0 (A–B)	1.6, 2.1; blue-white stars
			90.6 (A–C)	mag. 5 component
β Cyg	19^h 31^m	+27° 57'	34.4	3.0, 5.3; yellow and blue
γ Del	20^h 47^m	+16° 08'	9.2	4.0, 5.0; yellow stars
ϑ Eri	02^h 58^m	−40° 18'	8.3	3.4, 4.4; blue-white stars
α Gem	07^h 35^m	+31° 54'	3.7	2.0, 3.0; blue-white stars
δ^1, δ^2 Gru	22^h 29^m	−43° 30'	wide optical	n.e., 4.0, 4.1; yellow and red giants
μ^1, μ^2 Gru	22^h 16^m	−41° 21'	wide optical	n.e., 4.8 and 5.1; yellow giants
α Her	17^h 15^m	+14° 24'	4.8	"Ras Algethi": 3.5, 5.1
π^1, π^2 Hyi	02^h 14^m	−67° 50'	wide optical	binocular pair; 5.6 and 5.7
γ Leo	10^h 19^m	+19° 51'	4.5	2.0, 3.1; yellow stars
ε Lyr	18^h 44^m	+39° 40'	2.7 (A–B)	"Double Double": 4.7, 5.8; 4.9, 5.2
			210.5 (AB–CD)	
			2.5 (C–D)	
β Mon	06^h 29^m	−07° 02'	7.2	4.7, 5.2; white stars
ϑ Mus	13^h 08^m	−65° 18'	5.3	5.4, 7.5
β Ori	05^h 15^m	−08° 12'	9.1	0.3, 10.4; difficult
η Per	02^h 51^m	+55° 53'	28.7	3.8, 8.5; orange and blue
ζ^1, ζ^2 Ret	03^h 18^m	−62° 35'	wide optical	n.e., 5.2 and 5.5, yellow stars
β Sco	16^h 05^m	−19° 48'	13.6 (A–C)	2.6, 4.9; blue-white
δ Ser	15^h 35^m	+10° 32'	4.1	3.8, 4.8; white stars
β Tuc	00^h 32^m	−62° 57'	27.0 (A–C)	4.4, 4.5
ζ UMa	13^h 24^m	+54° 55'	wide optical	"Mizar": n.e., 2.1, 4.2
γ Vel	08^h 09^m	−47° 21'	41.4	1.8, 4.2; blue-white; quadruple system
γ Vir	12^h 42^m	−01° 27'	1.7	2.7, 2.8; yellow-white stars
γ Vol	07^h 09^m	−70° 30'	14.1	3.8, 5.7; yellowish stars
				(n.e = naked eye)

In true binaries the positions of the stars change as they orbit one another, but the brighter star is treated as fixed, and the relative position of the fainter is determined. The orbital periods may be extremely long, being measured in decades or even centuries. If measurements are made over a long interval the orbit can be drawn. The shape and size of this depend upon the orientation in space. At times some pairs may be easily split, but at others they close and become difficult objects. For this reason, the more detailed lists of doubles and multiples normally give the date of the latest measurements. Despite the large amount of highly accurate data returned by the Hipparcos satellite, for example, many periods and orbits remain poorly known, because observations have been made over just a small portion of an orbit. Such systems are therefore well worth studying.

Measuring double stars

The measurement of doubles really requires a long-focus telescope (refractors and catadioptrics are generally preferred). It would be a mistake, however, to assume that reflectors cannot be used – as is often said to be the case. Large numbers of double-star observations have been obtained with reflectors, and some of the greatest observers of the 20th century, who discovered many doubles, always used reflectors. In addition to a long focal ratio, a proper mounting and drive, and a micrometer are required. There are several different forms of micrometer, but the most readily understandable is the filar micrometer, which incorporates fixed and moving threads (still sometimes made from spider's web). The position of the moving threads is controlled by accurately cut screws, fitted with graduated drums from which the number of rotations (or fraction of a rotation) can be read. Once calibrated for any particular telescope, measurements of the changes in position of the threads can be converted into seconds and minutes of arc, to give the separation between the stellar components. It is normally possible to rotate the micrometer around the optical axis, so that an index, read against a graduated scale, shows the Position Angle of the line joining the two components, relative to the north point of the field. Because of the difficulty of obtaining precise values (and instrumental errors), averages have to be taken of many individual measurements. The relative complexity of obtaining measurements, and the specialized instrumental requirements, make this a very neglected field of observation.

Description of the various different types of micrometer, and the precise way in which each one is used, would take us beyond the scope of this book, so the reader is referred to the publications listed under "Further Reading." However, it may be noted that although the filar micrometer has been described, two other forms are generally more suitable for amateur use. These are the binocular micrometer, where an artificial star is projected into a second eyepiece, and the diffraction micrometer, which uses a coarse grid of parallel wires, placed over the main objective.

The phenomenon of diffraction can be used with reflecting telescopes in another way, to increase the visibility of close companions. The secondary mirror in a Newtonian or Cassegrain telescope is normally held in the center of the tube by a "spider" with three or four vanes. This produces diffraction spikes around bright stars, which may hinder the detection of faint companions. By fitting a mask to the tube (typically with six or eight straight sides), the diffraction pattern can be modified, helping to reveal close companions.

In many doubles the components are too close to be resolved directly through any telescope. However, when their spectra are examined, doubling of the lines shows that more than one star is involved. These spectroscopic binaries are very numerous, and some have been examined by the few dedicated amateurs who undertake spectroscopic work. So many binary systems exist that most stars are part of a double or multiple system. The Sun is most unusual in not having any stellar companion.

Clusters

Apart from the stars that occur in binary and multiple systems, many are found in distinct groups, known as clusters. There are two main types, the open clusters (often called galactic clusters) and the spherical globular clusters. The open clusters, in particular, cover a wide range of sizes, and are therefore observable with a wide range of instruments. In both cases, photographs or CCD images can record far more detail than is visible to the naked eye, but many amateurs still like to make drawings of clusters, which they can then compare with images obtained in other ways. Sometimes there are striking differences in what is recorded.

Open clusters

The open clusters are irregular in shape and are groups of stars that formed together from a single interstellar dust and gas cloud, and which thus have similar ages and compositions. They are mainly found in the spiral arms of our Galaxy and as a result are concentrated in the regions of the Milky Way. These clusters can vary greatly in the number of stars within them. Some may be difficult to distinguish from the surrounding star fields, appearing as slightly denser patches of stars. These are often old clusters that have gradually spread out because of

▲ Fig. 11.12 The Double Cluster in Perseus, with χ Per and its red stars on the left, and h Per on the right, nearer to Cassiopeia.

each star's individual motion and thus become less distinct. On the other hand, younger clusters, such as the Pleiades (M45), are often densely crowded and contain many hot, young stars. Some clusters have distinctive shapes, the most notable of these perhaps being Brocchi's Cluster, "The Coathanger," which (for once) does resemble the object after which it is named.

Depending on the object being examined, and the equipment available, it may not be possible to resolve an open cluster into individual stars, and all that may be visible is a fuzzy patch resembling nebulosity. However, many open clusters are embedded in emission nebulosity, which is energized by radiation from the stars. The Orion Nebula, illuminated by the stars of the Trapezium, is perhaps the most obvious example, but two other well-known ones are the Rosette Nebula, surrounding NGC 2244, in Monoceros, and the η Carinae Nebula in the southern Milky Way.

Because of their very nature, it is difficult to determine precise sizes or magnitudes for open clusters, and any listed values (such as those quoted in Table 11.5) should be taken as approximate, and merely a guide to a cluster's likely visibility. Similarly, stars that apparently belong to a cluster are often foreground stars that have no connection with the cluster itself. The most striking example of this is possibly Aldebaran, α Tauri, which is not a member of the prominent Hyades cluster. Although stars in a cluster have generally been born at about the same time, in some clusters the more massive stars have had time to evolve, passing from young blue stars to yellow, orange, or even red objects. Such stars often form a striking contrast with others in the cluster. The most notable example is probably NGC 4755, the Jewel Box, in the southern constellation of Crux, but there are many others, including the Double Cluster, h and χ Persei, both of which contain distinctly colored stars (Fig. 11.12).

TABLE 11.5: OPEN CLUSTERS						
Object	Const.	RA	Dec.	m	dia.	Name or notes
M6	Sco	$17^h 40^m$	$-32°$ 13'	4.6	20'	
M7	Sco	$17^h 54^m$	$-34°$ 49'	3.3	80'	
M11	Sct	$18^h 51^m$	$-06°$ 16'	6.3	–	Wild Duck
M16	Ser	$18^h 19^m$	$-13°$ 47'	6.4	–	in Eagle Nebula
M23	Sgr	$17^h 57^m$	$-19°$ 01'	5.9	30'	
M25	Sgr	$18^h 32^m$	$-19°$ 15'	6.2	30'	
M34	Per	$02^h 42^m$	$+42°$ 47'	5.8	25'	
M35	Gem	$06^h 09^m$	$+24°$ 20'	5.6	25'	
M36	Aur	$05^h 36^m$	$+34°$ 08'	6.5	10'	
M37	Aur	$05^h 52^m$	$+32°$ 33'	6.2	15'	
M38	Aur	$05^h 28^m$	$+35°$ 50'	6.8	15'	
M39	Cyg	$21^h 32^m$	$+48°$ 26'	5.3	30'	
M41	CMa	$06^h 46^m$	$-20°$ 44'	5.0	40'	
M44	Cnc	$08^h 40^m$	$+19°$ 59'	3.7	70'	Praesepe
M45	Tau	$03^h 47^m$	$+24°$ 07'	1.6	120'	Pleiades
M46	Pup	$07^h 42^m$	$-14°$ 48'	6.6	20'	
M47	Pup	$07^h 37^m$	$-14°$ 29'	4.3	25'	naked-eye object
M48	Hya	$08^h 14^m$	$-05°$ 48'	5.5	30'	
M50	Mon	$07^h 03^m$	$-08°$ 20'	7.2	15'	
M52	Cas	$23^h 24^m$	$+61°$ 35'	8.2	16'	
M67	Cnc	$08^h 50^m$	$+11°$ 49'	7.4	25'	
M93	Pup	$07^h 44^m$	$-23°$ 51'	6.5	10'	
M103	Cas	$01^h 33^m$	$+60°$ 42'	6.9	6'	red star in field
NGC 457	Cas	$01^h 19^m$	$+58°$ 20'	5.1	20'	
NGC 752	And	$01^h 57^m$	$+37°$ 41'	6.6	75'	
NGC 869	Per	$02^h 19^m$	$+57°$ 09'	4.3	18'	h Per
NGC 884	Per	$02^h 22^m$	$+57°$ 07'	4.4	18'	χ Per
NGC 1502	Cam	$04^h 08^m$	$+62°$ 20'	4.1	20'	
NGC 1981	Ori	$05^h 35^m$	$-04°$ 26'	4.2	28'	
NGC 2169	Ori	$06^h 08^m$	$+13°$ 57'	7.0	6'	
NGC 2232	Mon	$06^h 27^m$	$-04°$ 45'	4.2	45'	
NGC 2244	Mon	$06^h 32^m$	$+04°$ 52'	5.2	30'	in Rosette Nebula
NGC 2264	Mon	$06^h 41^m$	$+09°$ 53'	4.1	40'	
NGC 2451	Pup	$07^h 46^m$	$-37°$ 57'	3.7	50'	
NGC 2477	Pup	$07^h 52^m$	$-38°$ 32'	5.7	20'	
NGC 2516	Car	$07^h 58^m$	$-60°$ 51'	3.3	22'	
NGC 2547	Vel	$08^h 11^m$	$-49°$ 15	5.0	25'	
NGC 3532	Car	$11^h 07^m$	$-58°$ 39'	3.4	50'	
NGC 3766	Cen	$11^h 36^m$	$-61°$ 35'	4.6	15'	
NGC 4755	Cru	$12^h 54^m$	$-60°$ 19'	5.2	–	Jewel Box near red κ
NGC 5460	Cen	$14^h 08^m$	$-48°$ 18'	6.1	35'	

TABLE 11.5: OPEN CLUSTERS (cont.)						
Object	Const.	RA	Dec.	m	dia.	Name or notes
NGC 5822	Lup	15^h 05^m	−54° 19'	6.5	35'	
NGC 6025	TrA	16^h 04^m	−60° 29'	6.0	15'	bright cluster
NGC 6087	Nor	16^h 13^m	−54° 11'	6.5	15'	
NGC 6231	Sco	16^h 54^m	41° 48'	3.4	−	
NGC 6633	Oph	18^h 28^m	+06° 34'	5.6	20'	
NGC 6709	Aql	18^h 52^m	+10° 21'	7.4	15'	
IC 2391	Vel	08^h 40^m	−53° 03'	2.6	60'	
IC 2395	Vel	08^h 41^m	−48° 11'	4.6	17'	
IC 2602	Car	10^h 43^m	−64° 23'	1.6	100'	Southern Pleiades
IC 4665	Oph	17^h 46^m	+05° 43'	4.2		
Melotte 111	Com	11^h 25^m	+26° 10'	4.0	120'	Coma Star Cluster
Cr 399	Vul	19^h 25^m	+20° 11'	3.6	−	Brocchi's Cluster, Coathanger

Globular clusters

Globular clusters are dense spherical concentrations of stars which may contain thousands or even millions of individual stars. They are very old and were formed very early in the history of the Galaxy itself, long before any heavy elements had been produced by nuclear fusion within stars and redistributed into space by their explosion. Unlike open clusters they are not found in the spiral arms, but are concentrated around the center of the Galaxy in Sagittarius. They are also found far out in the galactic halo. Compared with open clusters, fewer globular clusters are observable with amateur-sized equipment, and indeed most of the brighter ones – with the major exception of ω Centauri and 47 Tucanae in the southern hemisphere

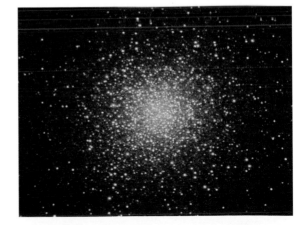

▶ Fig. 11.13 M13 in Hercules, the finest globular cluster in the northern sky, photographed by Ed Grafton.

TABLE 11.6: BRIGHT GLOBULAR CLUSTERS						
Object	Const.	RA	Dec.	m	dia.	Name and notes
M2	Aqr	$21^h 34^m$	$-00°$ 49'	9.2	6'	
M3	CVn	$13^h 42^m$	$+28°$ 23'	6.2	16'	
M4	Sco	$16^h 24^m$	$-26°$ 32'	5.6	26'	
M5	Ser	$15^h 19^m$	$+02°$ 05'	5.7	17'	
M9	Oph	$17^h 19^m$	$-18°$ 31'	7.7	9'	
M10	Oph	$16^h 57^m$	$-04°$ 06'	6.6	15'	
M12	Oph	$16^h 47^m$	$-01°$ 57'	6.7	15'	
M13	Her	$16^h 42^m$	$+36°$ 28'	5.8	17'	finest northern globular
M14	Oph	$17^h 38^m$	$-03°$ 15'	7.6	12'	
M15	Peg	$21^h 30^m$	$+12°$ 10'	6.2	12'	
M19	Oph	$17^h 03^m$	$-26°$ 16'	6.8	14'	
M22	Sgr	$18^h 36^m$	$-23°$ 54'	5.1	24'	3rd brightest globular
M53	Com	$13^h 13^m$	$+18°$ 10'	7.6	13'	
M62	Oph	$17^h 01^m$	$-30°$ 07'	6.5	14'	
M71	Sge	$19^h 54^m$	$+18°$ 47'	8.2	7'	
M92	Her	$17^h 17^m$	$+43°$ 08'	6.4	11'	
NGC 104	Tuc	$00^h 24^m$	$-72°$ 03'	4.0	31'	47 Tucanae, 2nd brightest globular
NGC 362	Tuc	$01^h 03^m$	$-70°$ 49'	6.4	13'	
NGC 1261	Hor	$03^h 12^m$	$-55°$ 12'	8.2	7'	
NGC 4833	Mus	$13^h 00^m$	$-70°$ 53'	6.9	14'	
NGC 5139	Cen	$13^h 27^m$	$-47°$ 29'	3.7	36'	ω Centauri, brightest globular
NGC 5986	Lup	$15^h 46^m$	$-37°$ 47'	7.5	10'	
NGC 6397	Ara	$17^h 41^m$	$-53°$ 40'	5.7	26'	
NGC 6541	CrA	$18^h 08^m$	$-43°$ 42'	6.3	13'	
NGC 6752	Pav	$19^h 11^m$	$-59°$ 58'	5.4	20'	

– were noted by Messier in his original catalog. The list in Table 11.6 gives some of the brighter and more obvious objects. As with open clusters, the details of magnitude and diameter should be treated with caution.

Once again, the question of resolution comes into play. The differences between globular clusters become most apparent when larger instruments are used. With binoculars or small telescopes many globulars simply appear as hazy, circular patches, all of which look more-or-less alike. With better resolution, however, differences become apparent in the degree to which they are condensed toward the center, and the appearance of specific lines of stars or dark lanes.

DEEP-SKY OBJECTS

Nebulae

A large amount of gas and dust exists within the Galaxy and this is often concentrated into the dense clouds known as nebulae. These may be divided into several categories.

Dark nebulae

In the dark nebulae thick masses of dust block the light from distant stars. There are several regions like this along the Milky Way, such as the Great Rift in Cygnus, the Coalsack in Crux, and the two dark lanes that form a prominent "V" shape near the galactic center in Sagittarius (Fig. 12.1). Apart from these dense clouds, however, there are many fainter ones, some of which are very difficult to see, and which wind their way across the general star clouds of the Galaxy. Low magnifications and excellent conditions are needed to reveal some of these faint dark lanes. Because gas and dust often occur together, many dark nebulae are associated with bright emission nebulae, or are seen outlined against glowing gas. The "Fish Mouth" which breaks into the Orion Nebula (M42) is one example, and another consists of the three dark lanes that give the Trifid Nebula (M20) its name.

Reflection nebulae

Dust is also responsible for the reflection nebulae. There are not very many that can be seen by visual observers, but some of the nebulosity in the Pleiades, for example, may be glimpsed under good conditions. Reflection nebulae are notoriously difficult to observe, mainly because the bright star or stars illuminating the nebula frequently overpower the faint nebulosity. This is one reason why such nebulae are often easy to see on long-exposure photographs, but remain elusive to visual observers. Another reason is that the amount of light reflected by the nebulosity drops rapidly with increasing distance from the star, so such

▶ Fig. 12.1 Prominent dark dust lanes occur near the galactic center in Sagittarius.

◀ *Fig. 12.2 The Rho Ophiuchi complex of molecular clouds, photographed by Stephen Pitt.*

nebulae tend to be small in size. Any scattered light within the telescope will also cause a loss of contrast, so in this field in particular perfectly clean optical surfaces – although always desirable – are almost essential.

Many reflection nebulae, such as the Pleiades nebulosity, appear blue; the clouds of dust reflect light from hot, young stars lying in front of them. A few others, such as the extensive nebulosity near Antares in Scorpius, are yellow or red, depending on the colors of the stars that illuminate them. The area around Antares is part of a larger complex, known as the ρ Ophiuchi molecular cloud (Fig.12.2), which exhibits a whole range of reflected colors from blue to red. Even though gas may be present in reflection nebulae, only in a few cases do the stars provide enough energy for it to glow.

Emission nebulae

In the emission nebulae, ultraviolet light from stars within them is absorbed and re-emitted at visible wavelengths. Visually, they appear greenish, because the eye is most sensitive to light in that range, and this is where light emitted by oxygen occurs. Photography, however, tends to show the red of glowing hydrogen. The most notable example is the famous Orion Nebula (M42). This is just visible to the naked eye as a hazy "star" in Orion's "sword," and appears as a distinct pink spot of light even on wide-field photographs taken with stationary or driven cameras. Telescopes reveal a vast glowing cloud of gas, surrounding the hot, young stars of the "Trapezium" (ϑ Orionis), which are the source of the radiation that excites the nebula (Fig. 12.3). Emission and dark nebulae are often the regions where new stars are being formed.

In contrast to reflection nebulae, emission nebulae are usually large, and thus much easier to detect. Generally, a low magnification is required. Low-power, large-aperture telescopes – the so-called "rich-field telescopes" – are ideal for this work, reflectors being best because of their lack of chromatic aberration. With large telescopes, colors other than green may become detectable visually – the Orion

Object	Const.	RA	Dec.	m	Name or notes
		TABLE 12.1: BRIGHT GASEOUS NEBULAE			
M8	Sgr	18ʰ 04ᵐ	−24° 23′	6.0	Lagoon Nebula
M17	Sgr	18ʰ 21ᵐ	−16° 11′	7.0	Omega Nebula
M42	Ori	05ʰ 35ᵐ	−05° 27′	4.0	Orion Nebula
NGC 2070	Dor	05ʰ 39ᵐ	−69° 10′	–	30 Dor, Tarantula, in LMC
NGC 2237	Mon	06ʰ 32ᵐ	+04° 52′	4.8	Rosette Nebula
NGC 3372	Car	10ʰ 54ᵐ	−59° 55′	–	η Carinae Nebula
NGC 6960	Cyg	20ʰ 46ᵐ	+30° 43′	–	Veil Nebula
NGC 6992	Cyg	20ʰ 56ᵐ	+31° 43′	–	Veil Nebula
NGC 7000	Cyg	20ʰ 59ᵐ	+44° 20′	–	North America

Nebula, for example, appears greenish in the center and takes on a reddish tinge toward the outside.

Because the light from the gas is emitted at very specific wavelengths, emission nebulae (and also planetary nebulae, described next) benefit from the use of "nebular filters" (also called "light-pollution filters"). Light from the most common sources of pollution – generally sodium- and mercury-vapor lamps – is blocked, while the light from the gaseous emission lines is transmitted through the filter. Use of an appropriate visual or photographic filter may dramatically improve the visibility of a nebula, even in cases where there is only slight light pollution.

Planetary nebulae

Another form of nebula is created when an evolving star reaches the end of its red-giant phase, and sheds a shell or shells of gas. These planetary nebulae (so named from their appearance) are also glowing, and sometimes (with large telescopes) the small, hot stellar remnant can be seen in the very center. In general, the diameters of planetary nebulae are very small, which means that few are readily seen with amateur equipment (Table 12.2). The most conspicuous is M27 in

▶ *Fig. 12.3 The Orion Nebula, M42, is the most prominent emission nebula, and visible to the naked eye. Photograph by Steve Massey.*

TABLE 12.2: BRIGHT PLANETARY NEBULAE					
Object	Const.	RA	Dec.	m	Name or notes
M27	Vul	$19^h\,59^m.6$	$+22°\,43'$	7.4	Dumbbell
M57	Lyr	$18^h\,53^m.6$	$+33°\,02'$	8.8	Ring
NGC 3242	Hya	$10^h\,24^m.8$	$-18°\,38'$	7.8	Ghost of Jupiter
NGC 7293	Aqr	$22^h\,29^m.8$	$+07°\,58'$	–	Helix

Vulpecula (Fig. 12.4), which may be seen in binoculars as a faint glow. A moderate-sized telescope is required to detect M57, the famous Ring Nebula in Lyra, as a vaguely elliptical hazy patch, and much larger telescopes are needed to reveal the dark center.

As with clusters and galaxies, the magnitudes often given for planetary nebulae in various lists (including those in the table above) must be treated with considerable caution. Ease of visibility depends on many different factors. Because, like emission nebulae, planetary nebulae emit light at very specific wavelengths, visibility may be improved by the use of nebular filters.

Tiny planetary nebulae can also be detected by a different method – by spreading their light into a spectrum using a spectroscope or diffraction grating, as described below. Instead of the continuous spectrum shown by the vast majority of stars, planetaries display just a few bright lines, and thus stand out in comparison with neighboring stars. This is particularly striking when a planetary nebula occurs within a globular cluster – it is immediately obvious that it is a different type of object.

Occasionally it may be possible to obtain a relatively old-fashioned accessory known as a direct-vision spectroscope. This consists of a small train of prisms held between the eye and the telescope's eyepiece. The dispersion of the prisms is arranged so that one does obtain a

▶ Fig 12.5 The North America Nebula (left) and the Pelican Nebula (right).

▼ Fig. 12.4 The planetary nebula M27.

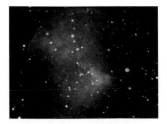

▶ *Fig. 12.6 The expanding gas clouds in M1 in Taurus, a supernova remnant, have a highly complex structure. Photograph by Gordon Rogers.*

direct view of the field. An alternative is to hold a glass prism or a diffraction grating between the eye and the eyepiece, but in this case the light is turned through a considerable angle and it becomes much more difficult to identify the field.

Plastic diffraction gratings are readily available for a small cost – often supplied in 35 mm slide mounts. Quite apart from their use for detecting planetary nebulae, it is instructive to examine various artificial lights through them and thus determine their differences. When used on stars, the distinct dark bands in the spectra of very cool stars are often detectable. A few amateurs do carry out serious spectroscopic work, but discussion of this topic is beyond the scope of this book.

Supernova remnants

Nebulae that are superficially similar in appearance to planetary nebulae arise when very massive stars explode as supernovae. In this process, the stars are disrupted and their material is ejected into space, including the heavy elements formed within them at the instant of the explosion. These elements may later be incorporated into new stars and planets. The material is ejected into space at such high velocities that when it collides with pre-existing interstellar material, intense shock-waves are produced, and the energy released causes the gases to glow. (Much of the energy is actually radiated in X-rays, rather than in the visible spectrum.) Such nebulae are known as supernova remnants (SNR), and the Crab Nebula in Taurus (M1, Fig. 12.6) and the Veil Nebula, part of the enormous Cygnus Loop, are the most easily seen in amateur-sized telescopes (Table 12.3).

TABLE 12.3: SUPERNOVA REMNANTS					
Object	Const.	RA	Dec.	m	Name or notes
M1	Tau	05^h 34^m.5	+22° 01'	8.4	Crab Nebula
NGC 6960	Cyg	20^h 45^m.7	+30° 43'	–	Veil Nebula
NGC 6992	Cyg	20^h 56^m.4	+31° 43'	–	Veil Nebula

The Galaxy

The star clouds of the Milky Way run right around the sky. They are least distinct in the region of Gemini, Orion, and Auriga, where they may be seen on only the darkest nights. In the other half of the sky, however, between Cygnus in the north and Carina in the south, a clear night will reveal densely packed clouds of stars. In some areas the constellation patterns formed by the brightest stars may even be difficult to pick out from the brilliant background. It is well worth scanning the Milky Way with low magnification, such as that provided by old-fashioned opera glasses or low-power binoculars. Dark "rifts" show the presence of dense clouds of dust (the dark nebulae mentioned earlier), absorbing the light from the stars beyond them. All these clouds of stars and dust, as well as the young open clusters, mark the plane of the spiral arms and the disk of our Galaxy. The diameter of this vast, thin disk is about 30 kiloparsecs or roughly 100,000 light-years.

Wide-angle photographs sometimes give an indication of our Galaxy's appearance as a disk with a central bulge, which lies in the general direction of the constellation of Sagittarius. This lens-shaped bulge is the galactic nucleus, a flattened ball of old, reddish stars, very distinct from the young, blue stars of the spiral arms. It surrounds the galactic center, which lies far away in Sagittarius, close to the border with Ophiuchus (Fig. 12.7). In visible light it is hidden from us by the dense clouds of dust in the galactic plane, but γ-ray, X-ray, infrared, and

radio observations reveal it to be the site of gigantic, swirling clouds of gas, huge star clusters, and in the very center, a massive black hole. Our Sun and Solar System lie well out toward the edge of the Galaxy, about 10 kiloparsecs (32,000 light-years) from the center.

But there is yet another, even larger, but far less distinct, part of our Galaxy. This is the galactic halo, a vast sphere of space, stretching out beyond even the galactic disk, its indefinite boundaries

◄ Fig. 12.7 The Milky Way in Sagittarius. Part of the galactic bulge is clearly seen below the central plane of the Galaxy, with its dust lanes and emission nebulae. Photograph by Stephen Pitt.

► Fig. 12.8 The great
Andromeda Galaxy, M31,
with its satellite galaxies,
M32 (above left) and
M110 (below right),
photographed by
Rob Gendler.

marked by far-distant globular clusters. It also contains an unknown amount of gas, a thinly scattered collection of faint, individual stars, and a large amount of dark matter, whose presence is deduced from the motions of the Galaxy's visible components.

How would the Galaxy appear from outside? Probably like the nearest large galaxy M31 (the Andromeda Galaxy – Fig. 12.8), the more distant M81 in Ursa Major, or perhaps even somewhat like the famous, nearly edge-on galaxy M104 (the "Sombrero") in Virgo, with its thin, dark band of obscuring dust. It is possible that our Galaxy has a short bar across its center, like the barred spirals described in the next section.

Galaxies

The galaxies, isolated stellar systems, are very varied in size and shape. Some are small and irregular, like the Small Magellanic Cloud (SMC), which merely appears like a detached part of the Milky Way. Others, such as the Large Magellanic Cloud (LMC), the nearest system to our own, are rather larger and show some slight organized structure. This is not readily apparent with amateur-sized equipment and techniques, however, with which the LMC seems relatively formless.

Other galaxies may be broadly divided into two types: the spirals and the ellipticals, more properly called ellipsoidal galaxies. Spiral galaxies (designated by the letter S) are like our own – flattened systems with a disk, central nucleus, and hot, young stars. A few rare ones (the S0 galaxies) have no spiral arms, but in all the others the structure can usually be seen if the face of the disk is turned toward us. The loosely wound arms of Sc galaxies like M33 in Triangulum are not so easy to make out in a telescope as are the closer arms of Sb spirals such as the magnificent M81 in Ursa Major. The tighter Sa galaxies may appear almost structureless in amateur-sized telescopes, but show details in long-exposure photographs. In the barred spiral galaxies (SB) the arms grow out of the ends of a distinct bar across the center. M95 in Leo, an SBb galaxy, is one example.

The smooth, elliptical galaxies contain only old stars, and very little gas and dust. They are classified by the letter E followed by a number to indicate the amount of flattening. The E0 galaxies, like M87, the giant elliptical in Virgo, appear completely spherical. M49, also in Virgo, is more flattened and is an E4 galaxy. The most extreme cases, the E7 galaxies, are very rare. They may appear almost rectangular and like an edge-on spiral galaxy. The smallest, dwarf ellipsoidal galaxies may only have one-millionth of the mass of our Galaxy, while the large, giant ellipsoidals (like M87) may be more than 100 times as massive.

Because the plane of our Galaxy is so heavily obscured by dust, we see most galaxies when looking out toward the Galactic Poles. This is especially the case in the northern hemisphere, where galaxies cluster thickly in the constellation of Coma Berenices and in nearby Virgo. Some of the most notable visual (and photographic) objects are given in Table 12.4, although the two Magellanic Clouds are excluded, because they are large, readily detectable, and any magnitude or position would be purely of academic interest.

Yet again, listed magnitude values give only an approximate idea of a galaxy's visibility and numerous other factors come into play, including surface brightness, angular size, inclination, and galaxy type. An edge-on spiral galaxy presents a distinct contrast to the background,

TABLE 12.4: BRIGHT GALAXIES					
Object	Const.	RA	Dec.	m	Name and notes
M31	And	$00^h 43^m$	$+41° 16'$	3.5	Andromeda Galaxy
M32	And	$00^h 47^m$	$+40° 52'$	8.2	companion to M31
M33	Tri	$01^h 34^m$	$+30° 40'$	5.7	
M51	CVn	$13^h 30^m$	$+47° 15'$	8.4	Whirlpool Galaxy
M64	Com	$12^h 57^m$	$+21° 40'$	8.5	Blackeye Galaxy
M65	Leo	$11^h 19^m$	$+13° 05'$	9.3	
M66	Leo	$11^h 20^m$	$+12° 59'$	9.0	
M74	Psc	$01^h 37^m$	$+15° 47'$	9.2	
M77	Cet	$02^h 43^m$	$00° 00'$	8.8	
M81	UMa	$09^h 56^m$	$+69° 03'$	6.9	
M82	UMa	$09^h 56^m$	$+69° 40'$	8.4	Cigar Galaxy
M83	Hya	$13^h 37^m$	$-29° 52'$	7.6	
M101	UMa	$14^h 03^m$	$+54° 20'$	7.7	Pinwheel Galaxy
M104	Vir	$12^h 40^m$	$-11° 37'$	8.3	Sombrero Galaxy
M110	And	$00^h 40^m$	$+41° 41'$	8.0	
NGC 253	Scl	$00^h 48^m$	$-25° 16'$	8.0	
NGC 2403	Cam	$07^h 37^m$	$+65° 35'$	8.4	
NGC 3115	Sex	$10^h 05^m$	$-07° 43'$	8.3	
NGC 5178	Cen	$13^h 26^m$	$-43° 01'$	7.6	Centaurus A

► *Fig. 12.9 The large, but faint, face-on spiral galaxy, M33 in Triangulum, photographed by Gordon Rogers.*

and may thus be easier to see than a galaxy with wide-open spiral arms that is face-on. It is low surface brightness, coupled with its large size, that makes M33 in Triangulum (Fig. 12.9), for example, so much more difficult to detect than M31. With face-on spirals, the rare SO type (which show no arms) and the Sa spirals are easier to detect than Sb objects, which have more widely spaced arms, and these in turn are easier to see than the even looser Sc galaxies.

Because of the low surface brightness of many galaxies, any increase in contrast is an advantage. Filters to remove light pollution will definitely help, but even under perfect conditions a change of magnification may be of assistance, and should be tried once the object's location has been determined.

It is, of course, in imaging galaxies and other deep-sky objects that CCD photography has come into its own in recent years, and where amateurs have achieved striking results. Deep-sky objects have one advantage over planets, comets, and other Solar System objects in that they are unchanging. Many individual images – even obtained over many different nights – can therefore be combined digitally to give a final image. Although a similar technique was (and still is) used with conventional film, the latter requires long exposures and extremely careful guiding. Although guiding remains critical, CCD imaging enables many short exposures to be used, thus minimizing guiding errors. Most of the striking images shown here have been obtained by combining many individual images.

Although detecting and imaging galaxies is rewarding, the most scientifically valuable work may be carried out by monitoring galaxies for supernova outbursts. A certain success was obtained in the past by visual and photographic observers, but CCD imaging has revolutionized amateurs' contribution to this field.

A supernova explosion has not been recorded in our own Galaxy since two events in 1572 and 1604, which were observed by the famous

▲ *Fig. 12.10 Pair of images of a galaxy without (top) and with a supernova.*

▲ *Fig. 12.11 CCD image of Gamma-Ray Burster 021004.*

astronomers Tycho Brahe and Kepler, respectively. But these outbursts are occasionally seen in other galaxies, and considerable numbers have been discovered by amateurs, particularly since the introduction of CCD cameras. These explosions are far greater than nova outbursts and the stars may rise by 20 magnitudes or more (Fig. 12.10). These stellar outbursts may be so spectacular that for a brief period a single supernova can far exceed the brightness of the whole of its galaxy, which may be a system of 100 thousand million normal stars.

Supernovae are of fundamental importance for determining extreme distances in the Universe, and also for a thorough understanding of the evolution of stars, galaxies, and the Universe itself. Even though there are automated programs that detect many supernovae, all such discoveries are of great importance, if only because they help to improve the statistical knowledge of different types of supernova explosion. As with novae, there are a number of supernova patrols, most notably the UK Nova/Supernova Patrol, coordinated by *The Astronomer*, which carries out rapid, but rigorous, checks on any "discovery" before passing information to the central clearing-house for astronomical discoveries. This patrol has achieved striking successes in recent times.

For many years, one of the greatest unsolved problems of astronomy was the nature of the events known as gamma-ray bursters. Here, intense short-lived bursts of gamma-rays were detected with no apparent association with any visible object. Recently, transient optical bursts have also been detected, indicating that these objects are extremely distant supernovae. Even in this field, where progress eluded professional astronomers with their giant telescopes for many years, amateurs have now succeeded in obtaining images with CCD equipment (Fig. 12.11), and are making a serious scientific contribution to the study of some of the most distant objects in the Universe.

RESOURCES

Books

Arnold, H. J. P. (2002), *Astrophotography*, Philip's, London

Bone, Neil (2007), *Aurora: Observing and Recording Nature's Spectacular Light Show*, Springer, London

Bone, Neil (1993), *Observer's Handbook: Meteors*, Philip's, London; & Sky Publ. Corp., Cambridge, Mass.

Burnham, Robert (1978), *Burnham's Celestial Handbook*, 3 vols., Dover Publications, New York

Cook, J., ed. (1999), *The Hatfield Photographic Lunar Atlas*, Springer-Verlag, Heidelberg

Dunlop, S. & Tirion, W. (2010), *Collins Night Sky*, HarperCollins, London

Dunlop, S., Rükl, A. & Tirion, W. (2005), *Collins Atlas of the Night Sky*, HarperCollins, London

Hoskin, M., ed. (1997), *Cambridge Illustrated History: Astronomy*, Cambridge University Press

Illingworth, Valerie & Clark, John O. E., ed. (2000), *Collins Dictionary of Astronomy*, 2nd edn, HarperCollins, London; & *Facts on File Dictionary of Astronomy*, 4th edn, Facts on File, New York

Karkoschka, E. (1999), *The Observer's Sky Atlas*, 2nd edn, Springer-Verlag, New York

Mobberley, M. (1995), *Astronomical Equipment for Amateurs*, Springer-Verlag, Heidelberg

Moore, P., ed. (2002), *Astronomy Encyclopedia*, Philip's, London

Moore, P. (2000), *Exploring the night sky with binoculars*, 4th edn, Cambridge University Press, Cambridge

Mullaney, James & Tirion, Wil (2011), *Cambridge Atlas of Herschel Objects*, Cambridge University Press, Cambridge

Mullaney, James & Tirion, Wil (2009), *Cambridge Double Star Atlas*, Cambridge University Press, Cambridge

Philip's Planisphere, Latitude 51.5°N, Philip's, London

Ridpath, Ian, ed. (2004), *Norton's Star Atlas*, 20th edn., Pi Press, New York

Ridpath, Ian, ed. (2007), *Oxford Dictionary of Astronomy*, 3rd edn, Oxford University Press, Oxford

Ridpath, Ian & Tirion, Wil (2007), *Collins Stars and Planets Guide*, HarperCollins, London

Rükl, Antonín (1990), *Hamlyn Atlas of the Moon*, Hamlyn, London & Astro Media Inc., Milwaukee

Scagell, Robin (2009), *Philip's Stargazing with a Telescope*, Philip's, London

Scagell, Robin & Frydman, David (2007), *Philip's Stargazing with Binoculars*, Philip's, London

Tirion, Wil (2001), *Bright Star Atlas*, Willmann Bell Inc., Richmond, Virginia

Tirion, Wil (2011), *Cambridge Star Atlas*, 4th edn, Cambridge University Press, Cambridge

Tirion, Wil & Sinnott, Roger (1998), *Sky Atlas 2000.0*, 2nd edn, Cambridge University Press, Cambridge; & Sky Publ. Corp., Cambridge, Mass.

Handbooks

British Astronomical Association, *Handbook* (published yearly)

Royal Astronomical Society of Canada, *Observer's Handbook* (published yearly)

Journals

Astronomy, Astro Media Corp., 21027 Crossroads Circle, P.O. Box 1612, Waukesha, WI 53187-1612, USA.
Website: http://www.astronomy.com

Astronomy Now, Pole Star Publications, P.O. Box 175, Tonbridge, Kent TN10 4QX, UK. Website: http://www.astronomynow.com

Sky & Telescope, Sky Publishing Corp., Cambridge, MA 02138-1200, USA. Website: http://www.skyandtelescope.com/

Societies

American Association of Variable Star Observers

25 Birch Street, Cambridge, Mass. 02138-1205, USA.
Website: http://www.aavso.org/
A society primarily concerned with the observation of variable stars, with an international membership and vast database of past observations.

Astronomical League

Executive Secretary: P.O. Box 43235, Martel, OH 43335, USA.
Website: http://www.astroleague.org/
An umbrella organization to which many local astronomical societies in the USA belong, and which is able to provide contact information.

British Astronomical Association

Burlington House, Piccadilly, London W1J 0DU.
Website: http://www.britastro.org/baa
The principal British organization for amateur astronomers (with some professional members), particularly for those interested in carrying out observational programs. Its membership is, however, worldwide. It publishes fully refereed, scientific papers and other material in its well-regarded *Journal*.

Federation of Astronomical Societies
Website: http://www.fedastro.org.uk/
An organization that is able to provide contact information for local astronomical societies in the United Kingdom.

Royal Astronomical Society
Burlington House, Piccadilly, London W1J 0BQ.
Website: http://www.ras.org.uk/
The premier astronomical society, with membership primarily drawn from professionals and experienced amateurs. It has an exceptional library and is a designated center for the retention of certain classes of astronomical data. Its publications are the standard medium for dissemination of astronomical research.

Royal Astronomical Society of Canada
136 Dupont Street, Toronto, Ontario, M5R 1V2, Canada.
Website: http://www.rasc.ca
The principal Canadian society, including both amateur and professional astronomers. It has various regional centers, and its *Journal* contains amateur and professional contributions.

Society for Popular Astronomy
Website: http://www.popastro.com
A society for astronomical beginners of all ages, which concentrates on increasing members' understanding and enjoyment, but which does have some observational programs. Its journal is entitled *Popular Astronomy*.

Software
Planetary, Lunar and Stellar Visibility (planetary and eclipse freeware): Alcyone Software, Germany.
  Website: http://www.alcyone.de
Redshift 7, Maris Multimedia (distributed by USM).
  Website: http://www.redshift-live.com/en/
Starry Night & Starry Night Pro, Sienna Software Inc., Toronto, Canada.
  Website: http://www.starrynight.com

Internet sources
There are numerous sites with information about all aspects of astronomy, and all of those given above have numerous links. Although many amateur sites are excellent, treat any statements and data with caution. The additional sites listed overleaf offer accurate information. Please note that the URLs may change. If so, use a good search engine, such as Google, to locate the information source.

Information

Auroral information Michigan Tech:
http://www.geo.mtu.edu/weather/aurora/

Comets Cometography: http://cometography.com/

Deep sky objects Saguaro Astronomy Club Database:
http://www.virtualcolony.com/sac/

Eclipses: http://eclipse.gsfc.nasa.gov/eclipse.html

Meteors Meteor Showers Online:
http://meteorshowersonline.com/

Moon (inc. Atlas) Inconstant Moon: http://www.inconstantmoon.com/

Planets Solar System Exploration:
http://solarsystem.nasa.gov/planets/index.cfm

Satellites (inc. *Space Station*)
Heavens Above: http://www.heavens-above.com/
Visual Satellite Observer's Home Page: http://www.satobs.org/

Star charts
National Geographic Chart (very small scale):
http://www.nationalgeographic.com/features/97/stars/chart/index.html
Sky & Telescope Interactive Chart:
http://www.skyandtelescope.com/observing/skychart/

What's Visible
Skyhound: http://observing.skyhound.com/
Skyview Cafe (requires Java-enabled browser):
http://www.skyviewcafe.com
Stargazer: http://www.outerbody.com/stargazer/
(choose "New View" from "File," click and drag to change)

Institutes and organizations

European Space Agency: http://www.esa.int/

International Dark Sky Association: http://www.darksky.org/

Jet Propulsion Laboratory: http://www.jpl.nasa.gov/

Lunar and Planetary Laboratory: http://www.lpl.arizona.edu/

National Aeronautics and Space Administration:
http://www.nasa.gov/centers/hq/home/index.html

Solar Data Analysis Center: http://umbra.gsfc.nasa.gov/

Space Telescope Science Institute: http://www.stsci.edu/public.html

UK Space Agency: http://www.ukspaceagency.bis.gov.uk/default.aspx

GLOSSARY

aphelion The point on an orbit that is farthest from the Sun.

apogee The point on its orbit at which the Moon is farthest from the Earth.

appulse The apparently close approach of two celestial objects, such as two planets, or a planet and star.

celestial latitude The angular distance of an object, measured perpendicular to the ecliptic.

celestial longitude The angular distance of an object, measured eastward along the ecliptic, with the zero point at the vernal equinox.

conjunction The point in time when two celestial objects have the same celestial longitude. In the case of the Sun and a planet, superior conjunction occurs when the planet lies behind the Sun (as seen from Earth). For Mercury and Venus, inferior conjunction occurs when they pass between the Sun and the Earth.

culmination The point at which a celestial body crosses the observer's meridian. This occurs twice a day, but both culminations are visible only for circumpolar objects. Upper culmination is the point closer to the zenith (that is, above the pole), and lower culmination that farther away from the zenith (below the pole).

cusp One of the tips of the crescent Moon, or any planetary body showing a crescent phase.

declination One coordinate in the equatorial system applied to the celestial sphere, and corresponding to latitude on Earth. It is measured in degrees north (+) or south (−) of the celestial equator (0°), reaching +90° and −90° at the celestial poles.

direct motion Motion from west to east on the sky, that is, in increasing right ascension.

direct rotation Rotation of a planet or satellite counterclockwise when looking down on the north pole.

ecliptic The apparent path of the Sun across the sky throughout the year. Also the plane of the Earth's orbit in space.

elongation The point at which an inferior planet has the greatest angular distance from the Sun, as seen from Earth.

following Lying to the east – that is, having a greater right ascension than the reference object.

limb The edge of the apparent disk of any planetary or satellite body against the background sky. Note that a dark limb may be invisible.

lunation A complete cycle of lunar phases, for example, from one New Moon to the next.

meridian The great circle passing through the north and south poles of a body and the observer's position; or the corresponding great circle on the celestial sphere that passes through the north and south celestial poles and also through the observer's zenith.

nadir The point on the celestial sphere directly beneath the observer's feet, opposite the zenith.

objective The principal image-forming element of any telescope, whether a mirror or lens.

opposition The point on a superior planet's orbit at which it is directly opposite the Sun in the sky, that is, when its RA differs by exactly 12 hours from that of the Sun.

penumbra 1. The outer region of a shadow cone, in which part of the light from the illuminating body is obscured, as in a partial eclipse.

2. The outer region of a sunspot, lighter than the umbra.

perigee The point on its orbit at which the Moon is closest to the Earth.

perihelion The point on an orbit that is closest to the Sun.

preceding Lying to the west – that is, having a lesser right ascension than the reference object.

precession The continuous alteration in the position of the vernal equinox (First Point of Aries) against the background sky, caused by the effects of the Sun and Moon on the orientation of the Earth's axis of rotation.

proper motion The motion of individual stars relative to the system of celestial coordinates, caused by the actual motion of the objects in space.

retrograde motion Motion from east to west on the sky – that is, in decreasing right ascension.

right ascension One coordinate in the equatorial system, corresponding to longitude on the Earth. It is measured on the celestial sphere in hours, minutes, and seconds, eastward from the vernal equinox.

terminator The line dividing the illuminated and non-illuminated portions of the Moon or other celestial body.

umbra 1. The central portion of a shadow cone, in which no direct light from the illuminating body is visible, as in a total eclipse.

2. The dark, central region of a sunspot.

vernal equinox The point at which the Sun, in its apparent motion along the ecliptic, crosses the celestial equator from south to north. Also known as the First Point of Aries.

zenith The point directly above the observer's head.

INDEX

Acknowledgments

Title Steve Massey; **7** Robin Scagell/Galaxy; **9** Chris Cook; **19** HJP Arnold/Sol Invictus; **24** Akira Fujii/DMI; **45** Stephen Pitt; **47t** Damian Peach; **47b** HJP Arnold/Sol Invictus; **49** Peter Grego; **53** Chris Cook; **56** Derek St Romaine; **58** Derek St Romaine; **59** Derek St Romaine; **63bl** Orion Optics/Galaxy; **63br** Robin Scagell/Galaxy; **65** Robin Scagell/Galaxy; **67** Peter Grego; **70** Robin Scagell/Galaxy; **74** Steve Edberg; **89bl** Robin Scagell/Galaxy; **89br** David Graham/Galaxy; **90** Peter Grego; **92** Chris Cook; **93** HJP Arnold/Sol Invictus; **94** Akira Fujii/DMI; **95** Steve Massey; **96** Stephen Pitt; **98** Chris Cook; **99** Chris Cook; **102** Dominic Cantin; **105** Neil Bone; **106** Stephen Pitt; **107** Chris Cook; **114** Dave Sewell; **116–17** Eckhard Slawik/Science Photo Library; **118** Peter Grego; **119t** Mike Goodall; **119b** Stephen Pitt; **122** HJP Arnold/Sol Invictus; **123** Mike Brown; **124** Mike Brown; **125bl** Mike Brown; **125br** Chris Cook; **126** Mike Brown; **128b** Akira Fujii/David Malin Images; **129** HJP Arnold/Sol Invictus; **132b** Paul Stephens; **134** Robin Scagell/Galaxy; **135** HJP Arnold/Sol Invictus; **137bl** Akira Fujii/David Malin Images; **137br and c** HJP Arnold/Sol Invictus; **142tl and tr** Richard McKim; **143bl and br** Richard McKim; **144** Richard McKim; **145** Donald C. Parker; **146** Richard McKim; **148–149** Tunç Tezel; **150tl** Richard McKim; **150tr** Tunç Tezel; **152** Bill Livingston/NOAO/AURA/NSF; **155** Steve Massey; **158** Damian Peach; **159t and c** Richard McKim; **160** Steve Massey; **161t** David Hanon; **161b** Damian Peach; **163** Naoyuki Kurita; **164c** Ed Grafton; **165** Naoyuki Kurita; **167t** Akira Fujii/David Malin Images; **167b** Thierry Legault/Eurelios; **168bl** A. Kelly; **168br** Akira Fujii/David Malin Images; **169** Richard McKim; **171** Ed Grafton; **176t** Akira Fujii/David Malin Images; **180** UC Regents/Lick Observatory; **181bl** Ed Grafton; **185** Ed Grafton; **187** Ed Grafton; **189** Stephen Pitt; **190** Stephen Pitt; **191** Steve Massey; **192bl** Ed Grafton; **192br** Stephen Pitt; **193** Gordon Rogers; **194** Stephen Pitt; **195** Rob Gendler; **197** Gordon Rogers; **198t, c, b** Mark Armstrong.

Star Maps by Wil Tirion; Moon Map by John Murray; artworks by Raymond Turvey and Wil Tirion; all © Philip's.